COLLEGIATE SCHOOL ENNISKILLEN			
Chem. DEPT.		NUMBER C/6B/5	
DATE ISSUED	DATE RETD	DATE ISSUED	DATE RETD

CHEMISTRY:
EXERCISES IN COMPREHENSION

H. Powell, B.Sc., A.R.I.C.
Head of Chemistry, Calday Grange Grammar School, West Kirby
R. J. Hateley, B.Sc.
Head of Chemistry, Wintringham Grammar School, Grimsby

THE ENGLISH UNIVERSITIES PRESS LIMITED

ISBN 0 340 16414 X

First printed 1968
Second edition 1972

Copyright © 1972 R. J. Hateley and H. Powell
All rights reserved. No part of this publication may be
reproduced or transmitted in any form or by any means,
electronic or mechanical, including photocopy, recording,
or any information storage and retrieval system, without
permission in writing from the publisher.

The English Universities Press Limited
St Paul's House, Warwick Lane, London EC4P 4AH

Printed and bound in Great Britain by
W. S. Cowell Ltd, Butter Market, Ipswich

CONTENTS

Books	6
Preface	7
A summary of terms used	9
Advice to the student	27

Questions

Physical

Acid – alkali titrations	28
Oxidation numbers and redox titrations	31
Molecules, moles, and equilibria	35
Colligative properties	39
Acids and acidity	43
Ionic equilibrium	47
Electrochemistry	51
Energetics	56
Phase equilibria	62
States of matter	69
Reaction kinetics	76

Inorganic

Periodic table	82
Group 1 M	86
Group 2 M	90
Group 3 M	94
Group 4 M	99
Group 5 M	103
Group 6 M	108
Group 7 M	113
Transition metals	117

Organic

Alkanes, alkenes, and alkynes	121
Alcohols and alkyl halides	125
Aldehydes, ketones, acids, and esters	129
Amines, amides, and cyanides	136
Aromatic compounds	140
Aromatic and aliphatic compounds	145
Qualitative and quantitative organic analysis	149

CONTENTS

Revision questions
    Physical   156
    Inorganic   171
    Organic   184

Answers and comments   196

Answers to Revision Questions   258

Periodic table

BOOKS

Students' guide to inorganic chemistry   J. BROCKINGTON, Butterworths
Introduction to physical chemistry   G. I. BROWN, Longmans
Inorganic chemistry   J. H. WHITE, University of London Press Ltd
Mechanism in organic chemistry   P. SYKES, Longmans
Chemical binding and structure   J. E. SPICE, Pergamon
Inorganic chemistry   J. MOELLER, Wiley
Carbon chemistry   M. G. BROWN, English Universities Press Ltd
Structural principles in inorganic compounds   W. E. ADDISON, Longmans
Modern chemistry   J. G. RAITT, Arnold
Organic / *Inorganic* / *Physical* } *chemistry*   C. W. WOOD AND A. K. HOLLIDAY, Butterworths
Physical chemistry   H. L. HEYS, Harrap
Chemistry: a unified approach   J. W. BUTTLE D. J. DANIELS AND P. J. BECKETT, Butterworths
Chemicals from petroleum   A. L. WADDAMS, Murray
Monographs for teachers   Royal Institute of Chemistry

PREFACE

This book has developed from our original efforts to design multiple-choice questions for sixth form tests as an alternative to the usual type of examination which, we felt, rarely tested understanding of the subject matter.

In the beginning the questions were arranged so that there was one correct answer out of a possible five. This still seems to be the best arrangement for test purposes and the revision exercises are all arranged so that there is one correct answer to each question.

The questions created much interest, particularly amongst the more able students, and it seemed that there might be a much greater use for them in helping the students to test themselves on different topics. Sets of twenty questions were given out to students as each topic was completed. The number of correct answers to each question varied from one to five and in some parts of the work there has been an effort to programme the answers so that the student was helped a little through the questions.

Most of the questions have been tried out with sixth forms and many of them have been the subject of much discussion. Whilst no claim is made that the problems are original, nevertheless the questions have not been taken from any other source, although we have obviously used a variety of sources for our facts and physical quantities.

It is hoped that the questions will be a stimulus to all students at A Level, no matter which syllabus they are following. The questions should prove useful to Ordinary National Certificate students in technical colleges, and scholarship candidates may find a number of questions which will give them food for thought.

In such a book of questions, which are designed to 'cross-link' the various topics as much as possible, it is very difficult to suit individual requirements and the arrangement of the topics is therefore our own compromise.

We would like to thank Mr. Derek Wareham and Mr. Douglas Little for their invaluable help and advice.

NOTES ON UNITS AND NAMES USED IN THE SECOND EDITION

We welcome the introduction of SI units and recognise their usefulness in calculations, but at the same time we realise that chemists will continue to use many of the old terms where these are more familiar or convenient.

We have deliberately left a few of the more common ones in the questions, particularly those with a practical bias. Here ml and cm^3 both appear, the former for liquid capacity and the latter for gaseous volumes.

Concentrations are in mol l^{-1}.

1 atmosphere pressure has been taken as 10^5 Nm^{-2} or 760mmHg.

°C has been retained for values involving such items as boiling and melting points, otherwise K is used.

A compromise has been made between the I.U.P.A.C. nomenclature for substances, and the names likely to be in school and textbook use for some years yet.

A SUMMARY OF THE TERMS USED IN THE QUESTIONS

This is intended as a brief description of the concepts used in the questions. Reference must be made to books such as those listed for a more detailed discussion.

1. Equations
Three types are used depending on which aspect of the reaction is being discussed.

a. Stoichiometric
These show the physical states of the reactants and the quantities concerned. The reaction between copper foil and silver nitrate solution would be:
$$Cu(s) + 2AgNO_3(aq) \rightarrow 2Ag(s) + Cu(NO_3)_2(aq)$$
1 mole 2 moles   2 moles 1 mole
(For the meaning of 'mole' see section 5)
No suggestions concerning the mechanism of such a reaction are implied.

b. Ionic
Ions common to both sides are omitted.
$$Cu(s) + 2Ag^+(aq) \rightarrow 2Ag(s) + Cu^{2+}(aq)$$

c. Electronic
This reaction also involves oxidation and reduction (see section 4).
The electronic half equations can be written:
$$2Ag^+(aq) + 2e^- \rightarrow 2Ag(s)$$
$$Cu(s) \rightarrow Cu^{2+}(aq) + 2e^-$$

2. Oxidation state and valency
The valency of manganese in $MnSO_4$ is 2 i.e. Mn^{2+}.
Valency is harder to appreciate in the permanganate ion MnO_4^- and here the concept of oxidation state (or number) is much more useful.
The oxidation number of an element can be conveniently defined as the actual charge on the ion of an element, or the formal charge on the element in a compound.

formula	ions		oxidation states		
H_2O	H^+	OH^-	H^I	O^{-II}	
Mn			Mn^0		
$MnCl_2$	Mn^{2+}	$2Cl^-$	Mn^{II}	Cl^{-I}	
MnO_2			Mn^{IV}	O^{-II}	
$KMnO_4$	K^+	MnO_4^-	Mn^{VII}	O^{-II}	K^I

Metals have positive oxidation states and as these are the most commonly used, the positive sign is usually omitted. The sum of the oxidation states in an uncharged molecule is zero.
The maximum oxidation state of an element, like the valency, is often the same as the group number or eight minus the group number.

A SUMMARY OF THE TERMS USED IN THE QUESTIONS

e.g. SO_4^{2-} : S^{VI} (Group 6M) O^{-II} (Group 6M)
    PO_4^{3-} : P^{V} (Group 5M) O^{-II} (Group 6M)
    CrO_4^{2-} : Cr^{VI} (Group 6T) O^{-II} (Group 6M)

This concept is most valuable in redox reactions (see section 4)
e.g. SO_3^{2-} : S^{IV} O^{-II}

This ion might be expected to be oxidised to SO_4^{2-} (i.e. S^{VI}) involving the transfer of 2 electrons

i.e. $S^{IV} \xrightarrow{\text{loss of } 2e^-} S^{VI}$

or $H_2O(l) + SO_3^{2-}(aq) \rightarrow SO_4^{2-}(aq) + 2H^+(aq) + 2e^-$

This is of course what happens with a suitable oxidising agent.
The oxidation state of an element can often with advantage be given in the name.

3. Systematic naming of compounds (*as recommended by I.U.P.A.C.*)

We have tried to encourage the systematic naming of inorganic and organic compounds as this is unambiguous and informative.

formula	trivial name	systematic name
$FeSO_4$	Ferrous sulphate	Iron(II) sulphate
$K_4Fe(CN)_6$	Potassium ferrocyanide	Potassium hexacyanoferrate(II)
$Cu(NH_3)_4SO_4$	Cuprammonium sulphate	Tetra-amminecopper(II) sulphate
$Na_3Co(NO_2)_6$	Sodium cobaltinitrite	Sodium hexanitrocobaltate(III)

4. Oxidation and reduction

Oxidation is conveniently defined as:
a a loss of one or more electrons from an atom, ion, or molecule;
b increase in the oxidation state of an element.

Reduction is exactly the opposite, and both processes must occur together (i.e. redox reactions). Electrons are always transferred to the oxidising agent from the reducing agent and the relevant electronic half equations are most conveniently balanced using oxidation states.

e.g. Oxidation of ferrous sulphate with potassium permanganate in acidic solution.

1. $FeSO_4(aq) \rightarrow Fe_2(SO_4)_3(aq)$
   or $Fe^{II} \rightarrow Fe^{III}$: $1e^-$ transferred
   i.e. $Fe^{2+}(aq) \rightarrow Fe^{3+}(aq) + e^-$
2. $KMnO_4(aq) \rightarrow MnSO_4(aq)$
   or $Mn^{VII} \rightarrow Mn^{II}$: $5e^-$ transferred
   i.e. $8H^+(aq) + MnO_4^-(aq) + 5e^- \rightarrow Mn^{2+}(aq) + 4H_2O(l)$

Balancing equation (i) $5Fe^{2+}(aq) \rightarrow 5Fe^{3+}(aq) + 5e^-$

As 1 mole of $MnO_4^-(aq)$ takes up 5 moles of electrons, the equivalent of potassium permanganate as an oxidising agent in acid solution is $\dfrac{\text{formula weight}}{5}$ (see section 6).

A SUMMARY OF THE TERMS USED IN THE QUESTIONS

5. The mole

The mole is the amount of substance which contains as many units as there are atoms in 12g of carbon 12. The units must be stated (e.g. atom, ion, molecule, etc.).
One mole contains Avogadro's constant (L) of units (i.e. 6×10^{23}).
All of the following constitute a mole (or mol):
23 g $Na_{(s)}$ or 1 g-atom
35·5 g $Cl_{(g)}$ or 1 g-atom
71 g $Cl_2{(g)}$ or 1 g-molecule
58·5 g $NaCl_{(s)}$ or 1 g-formula weight (consisting of 1 g-ion each of Na^+ and Cl^-)
6×10^{23} electrons (the number of electrons transferred from 23 g $Na_{(s)}$ to 35·5 g $Cl_{(g)}$ or 1 faraday of electricity)
A molar (M) solution of sodium chloride contains 58·5 g $NaCl_{(s)}$ dissolved in 1 litre of solution.
The mole avoids such phrases as 'g-equations worth' and 'g-formula weight' but care must be taken when using it: e.g. '1 mole of chlorine' usually means L molecules of chlorine gas ($Cl_2{(g)}$), but could be L atoms of chlorine ($Cl_{(g)}$).
Equations are best considered in terms of moles.

$$\text{e.g.} \quad Na_{(s)} \quad + \quad \tfrac{1}{2}Cl_2{(g)} \quad \rightarrow \quad NaCl_{(s)}$$
$$\quad 1 \text{ mole (23 g)} \quad \tfrac{1}{2} \text{ mole (35·5 g)} \quad 1 \text{ mole (58·5 g)}$$

6. Titrations

The results of titrations may be calculated using normalities and equivalents, or molarities and moles. We prefer the latter, but it will be some time before the concept of equivalent is discarded. There are problems given which can be worked out by either method, as in this example.
5·0 ml of hydrogen peroxide solution were diluted to 250 ml with distilled water and titrated in acid solution against potassium permanganate solution. 25 ml of diluted H_2O_2 required 28 ml N/10 $KMnO_4$ (or 28 ml 0.02M $KMnO_4$) for complete oxidation.
What was the concentration of the original hydrogen peroxide solution?

$$(H_2O_2 \rightarrow 2H^+ + O_2 + 2e^-$$
$$5e^- + 8H^+ + MnO_4^- \rightarrow Mn^{2+} + 4H_2O)$$

equivalents
25 ml diluted $H_2O_2 \equiv$ 28 ml N/10 $KMnO_4$
250 ml diluted $H_2O_2 \equiv$ 28 ml N $KMnO_4$

$$\equiv \frac{28}{1000} \text{ g-equivalent } KMnO_4$$

1 equivalent H_2O_2 reacts with 1 equivalent $KMnO_4$

molarities
25 ml diluted $H_2O_2 \equiv$ 28 ml 0.02M $KMnO_4$
250 ml diluted $H_2O_2 \equiv$ 28 ml 0.2M $KMnO_4$

$$\equiv \frac{28}{5000} \text{ mole } KMnO_4$$

2 moles $KMnO_4$ react with 5 moles H_2O_2

A SUMMARY OF THE TERMS USED IN THE QUESTIONS

equivalents

5 ml original H_2O_2 contain $\dfrac{28}{1000}$ g-equivalent H_2O_2

1 litre original H_2O_2 contains $\dfrac{28 \times 200}{1000}$ g-equivalent H_2O_2

$= 5\cdot6$ g-equivalent H_2O_2

From the above equation

1 g-equivalent $H_2O_2 = \dfrac{\text{Molar mass}}{2} = \dfrac{34}{2}$

$= 17$ g

Weight of H_2O_2 in 1 litre of original
solution $= 17 \times 5\cdot6$ g
$= \underline{95\cdot2\text{ g}}$

molarities

$\dfrac{28}{5000}$ mole $KMnO_4 \equiv \dfrac{5}{2} \times \dfrac{28}{5000}$ mole H_2O_2

$= \dfrac{14}{1000}$ mole H_2O_2

5 ml original H_2O_2 contains $\dfrac{14}{1000}$ mole H_2O_2

1 litre original H_2O_2 contains

$\dfrac{14}{1000} \times 200$ mole H_2O_2

$= 2\cdot8$ moles H_2O_2

Weight of H_2O_2 in 1 litre of original
solution $= 34 \times 2\cdot8$ g
$= \underline{95\cdot2\text{ g}}$

7. Energy considerations

Energy changes are fundamental to chemistry. We have tried to make use of a few simple concepts which are summarised here.

a. Thermochemical convention

Enthalpy or heat content of a system is designated by H. ΔH is the change in enthalpy and if ΔH is positive, heat is gained by the system (i.e. an endothermic reaction). If heat is lost from the system ΔH is negative.
Values quoted are for standard conditions: 760 mmHg and 298K.
It is important to state clearly the physical nature of the reactants.

	ΔH°
$H_2(g) + \tfrac{1}{2}O_2(g) \rightarrow H_2O(g)$:	-242 kJ
$H_2(g) + \tfrac{1}{2}O_2(g) \rightarrow H_2O(l)$:	-286 kJ
$H^+(aq) + OH^-(aq) \rightarrow H_2O(l)$:	-57 kJ

b. Ionisation energy

This is the energy required to remove an electron from an atom or ion in the gaseous state.
e.g. First ionisation energy: $Mg(g) \rightarrow Mg^+(g) + e^-$: $\Delta H^\circ = 736$ kJmol^{-1}
  Second ionisation energy: $Mg^+(g) \rightarrow Mg^{2+}(g) + e^-$: $\Delta H^\circ = 1450$ kJmol^{-1}
  Total energy required: $Mg(g) \rightarrow Mg^{2+}(g) + 2e^-$: $\Delta H^\circ = 2186$ kJmol^{-1}

An electrical potential of 7·6 volts applied to magnesium gas (in a valve) removes one electron from each atom. The electron acquires an energy of $7\cdot6 \times 1\cdot6 \times 10^{-19}$kJ as it is accelerated through

this potential and the total energy acquired by all the electrons removed from one mole (24 g) of magnesium amounts to $7\cdot 6 \times 1\cdot 6 \times 10^{-19} \times 6 \times 10^{23}$ joules
= 736 kJ

c. Electron affinity
The addition of an electron to a non-metal atom can liberate energy.
$Cl(g) + e^- \rightarrow Cl^-(g)$: $\Delta H = -370$ kJ mol^{-1}
This value is called the electron affinity.
When two electrons must be added then energy is *needed* as the second electron is added to a negatively charged ion.
$O(g) + 2e^- \rightarrow O^{2-}(g)$: $\Delta H = +627$ kJ mol^{-1}

d. Born-Haber cycle
The connection between these and other energy changes is best illustrated in diagrammatic form for the formation of sodium chloride from sodium and chlorine.
(All values are in kJmol^{-1} at 298K and 10^5Nm^{-2}.)

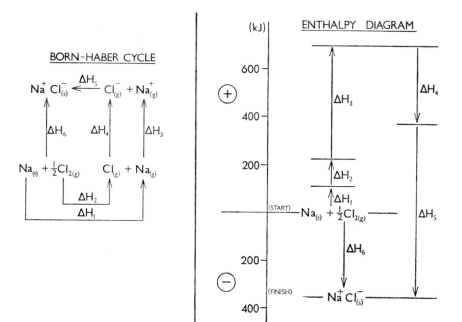

A SUMMARY OF THE TERMS USED IN THE QUESTIONS

Heat of atomisation of sodium　　　　　　　$\Delta H_1 = +109$ kJ
Heat of dissociation of chlorine　　　　　　$\Delta H_2 = +121$ kJ
Ionisation energy of sodium　　　　　　　　$\Delta H_3 = +494$ kJ
Electron affinity of chlorine　　　　　　　　$\Delta H_4 = -364$ kJ
Lattice (or crystal) energy of sodium chloride $\Delta H_5 = -771$ kJ
Heat of formation of sodium chloride　　　　$\Delta H_6 = -411$ kJ

e. Thermochemical problems
Calculations using Hess's Law are conveniently done by means of an enthalpy diagram.
e.g. Heat of combustion of carbon is 394 kJ (evolved).
　　Heat of combustion of sulphur is 297 kJ (evolved).
　　Heat of combustion of carbon disulphide is 1076 kJ (evolved).
What is the heat of formation (ΔH_f) of carbon disulphide?
(All values at 298K and $10^5 Nm^{-2}$.)

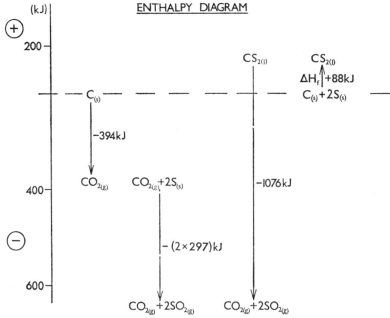

∴ Heat of formation of carbon disulphide is 88kJ mol^{-1} absorbed

A SUMMARY OF THE TERMS USED IN THE QUESTIONS

8. Electrode and redox potentials

The standard electrode potential is that electrical potential set up between an element and a 1 molar solution of its ions at 298K (designated E°).

The standard redox potential is the potential set up at 298K between an inert metal (platinum) and a solution of the oxidised and reduced species (e.g. Fe^{3+} and Fe^{2+}) in which the concentration of each is 1 M.

Typical values E° (volts)

$$Mg^{2+}(aq) + 2e^- \rightleftharpoons Mg(s) \quad\quad -2\cdot 37$$
$$Pb^{2+}(aq) + 2e^- \rightleftharpoons Pb(s) \quad\quad -0\cdot 13$$
$$2H^+(aq) + 2e^- \rightleftharpoons H_2(g) \quad\quad 0\cdot 00$$
$$Cl_2(aq) + 2e^- \rightleftharpoons 2Cl^-(aq) \quad\quad +1\cdot 40$$
$$MnO_4^-(aq) + 8H^+(aq) + 5e^- \rightleftharpoons Mn^{2+}(aq) + 4H_2O(l) \quad\quad +1\cdot 51$$

According to I.U.P.A.C. convention, the more positive the potential, the more likely is the reaction **as written above** to go from left to right.

The possibility of a reaction taking place can be predicted by constructing a series of cells.

a. Reaction between $Mg(s)$ and $Pb(NO_3)_2(aq)$.

$$Pt(s) : H_2(g) \,|\, H^+(aq) \,\|\, Mg^{2+}(aq) \,|\, Mg(s) \quad E^\circ = -2\cdot 37 \text{ V}$$
$$(760 \text{ mmHg}) \text{ (Molar)} \quad\quad \text{(Molar)}$$

                 Porous partition

Standard hydrogen electrode

The e.m.f. of this cell is the standard electrode potential for magnesium. As this is negative, then the reaction
$$H_2(g) + Mg^{2+}(aq) \rightarrow 2H^+(aq) + Mg(s)$$
DOES NOT occur.

b. Similarly for lead.
$$Pt(s) : H_2(g) \,|\, H^+(aq) \,\|\, Pb^{2+}(aq) \,|\, Pb(s) \quad E^\circ = -0\cdot 13 \text{ V}$$

Combining these two cells:

$$Mg(s) \,|\, Mg^{2+}(aq) \,\|\, H^+(aq) \,|\, H_2(g) : Pt(s) \quad - \text{ wire } - \quad Pt(s) : H_2(g) \,|\, H^+(aq) \,\|\, Pb^{2+}(aq) \,|\, Pb(s)$$
$$+2\cdot 37 \text{ V} \quad\quad\quad\quad\quad\quad\quad\quad\quad\quad\quad\quad\quad -0\cdot 13 \text{ V} \quad (\text{e.m.f.} = +2\cdot 24 \text{ V})$$
         (i.e. **a.** reversed)

or more simply: $Mg(s) \,|\, Mg^{2+}(aq) \,\|\, Pb^{2+}(aq) \,|\, Pb(s)$ e.m.f. $= +2\cdot 24$ V

As the e.m.f. of the cell is positive then the reaction
$$Mg(s) + Pb^{2+}(aq) \rightarrow Mg^{2+}(aq) + Pb(s)$$
DOES occur.

A SUMMARY OF THE TERMS USED IN THE QUESTIONS

This can be extended to other redox reactions.
e.g. $Cl_2(aq) + 2e^- \rightleftharpoons 2Cl^-(aq)$ $E° = +1·40$ V
$Mn^{2+}(aq) + 4H_2O(l) \rightleftharpoons MnO_4^-(aq) + 8H^+(aq) + 5e^-$ $E° = -1·5$ V
 SUM $= -0·1$ V

Therefore $Cl_2(aq)$ does NOT oxidise $Mn^{2+}(aq)$
but $MnO_4^-(aq)$ WILL oxidise $Cl^-(aq)$ in acid solution.
i.e. $MnO_4^-(aq) + 8H^+(aq) + 5e^- \rightleftharpoons Mn^{2+}(aq) + 4H_2O(l)$ $E° = +1·51$ V
$2Cl^-(aq) \rightleftharpoons Cl_2(aq) + 2e^-$ $E° = -1·40$ V
 SUM $= +0·11$ V

Such a series of redox potentials is obviously useful, but the following points should be noted.
(i) The numerical values can only be used strictly if the reaction being considered is carried out under the same conditions as those used to measure the standard redox potential (i.e. 25°C and molar concentration of all ions). In practice, however, most predictions from the values do work even if conditions are not the standard ones.
(ii) A positive sign for the e.m.f. of a cell means that the reaction will proceed, but there is no indication of how fast such a reaction will go.

9. Electronegativity

Electrons are not usually shared evenly between the atoms in a covalent bond. In $HF(g)$ the fluorine atom has a greater attraction for electrons than has hydrogen. In $HI(g)$ there is a less marked difference. The molecules are therefore dipoles and can be represented schematically.

H $\xrightarrow{+}$ F
$\delta+$ $\delta-$

Dipole moments can be measured (HF $= 1·9$, HI $= 0·4$ Debye units) and thus Pauling derived a table of electronegativity values which gives an indication of how uneven the electron sharing is.
Electronegativity values: H 2·1   O 3·5   F 4·0
Cl 3·0
Br 2·8
I 2·4

A difference in values between the two elements of about 1·7 or more indicates that the bond joining them is more ionic than covalent.

10. Shapes of molecules

The covalent hydrides CH_4, NH_3, H_2O and HCl may be conventionally written in one plane (only valency electrons being shown)

```
      H
      ··
H : C : H     H : N : H     H : O :     H : Cl :
      ··           ··           ··           ··
      H            H            H
```

16

A SUMMARY OF THE TERMS USED IN THE QUESTIONS

Such molecules are not planar. The electron pairs occupy space rather like balloons and mutually repel each other.

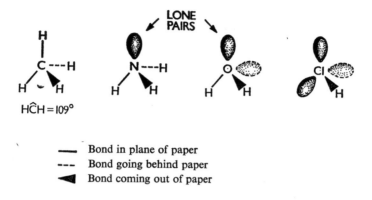

— Bond in plane of paper
--- Bond going behind paper
◀ Bond coming out of paper

The molecules thus have essentially a tetrahedral arrangement.
Where only 3 valency electrons are involved the molecule will be planar.

$$Cl\hat{B}Cl = 120°$$

This can be extended to double and treble bonds in organic molecules.

TETRAHEDRAL   PLANAR   LINEAR
($H\hat{C}H = 109°$)   ($H\hat{C}H = 120°$)   ($H\hat{C}C = 180°$)

A more sophisticated treatment involving hybridisation of s and p atomic orbitals gives the same overall picture.

A SUMMARY OF THE TERMS USED IN THE QUESTIONS

11. Ion hydration and hydrolysis

The water molecule by virtue of its shape and electron arrangement is readily attracted to metal ions, which are thus hydrated. One hydrated ion whose structure is known with certainty is $(Cr6H_2O)^{3+}$ with the water molecules arranged octahedrally. It is likely that the water in most hydrated ions is disposed in a similar symmetrical fashion.

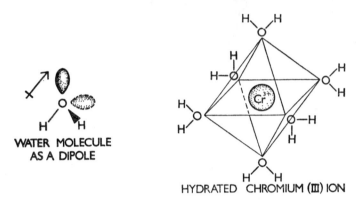

WATER MOLECULE AS A DIPOLE

HYDRATED CHROMIUM (III) ION

The attractive forces between the ion and the water molecules may be simply electrostatic in nature (as in $Na^+(aq)$), but where the ion is small with a large charge then hydrolysis can occur.

e.g. $\begin{pmatrix} H_2O & & OH_2 \\ H_2O & Al^{3+} & OH_2 \\ H_2O & & OH_2 \end{pmatrix} + H_2O \rightleftharpoons \begin{pmatrix} H_2O & & OH_2 \\ H_2O & Al^{3+} & OH_2 \\ H_2O & & OH^- \end{pmatrix} + H_3O^+$

Here $(Al6H_2O)^{3+}$ is acting as an acid on the Brønsted theory (i.e. a proton donor) and H_2O is acting as a base (proton acceptor). Thus aluminium salts give solutions which are acidic. For simplicity in the questions the hydrated proton is designated $H^+(aq)$ but occasionally H_3O^+ is used where this is more convenient.

12. Complexes

The replacement of water molecules in hydrated transition metal ions by other molecules or ions with lone pairs of electrons (e.g. NH_3, CN^-, Cl^-) gives complex ions. Water is held largely by ion-dipole attraction but with other ligands a much stronger bond is formed and the familiar chemical properties of the ion are lost.

$\begin{pmatrix} H_2O & & OH_2 \\ & Cu^{2+} & \\ H_2O & & OH_2 \end{pmatrix} \xrightarrow[H_2O]{NH_3} \begin{pmatrix} H_3N & & NH_3 \\ & Cu^{2+} & \\ H_3N & & NH_3 \end{pmatrix}$

Tetra-amminecopper(II) ion

A SUMMARY OF THE TERMS USED IN THE QUESTIONS

The 'rare-gas' electron configuration theory of valency can be applied to complexes with apparent success in some cases, but not in others.

e.g. Fe   2.8.14.2   Cu   2.8.17.2
      Fe^{2+} 2.8.14 Cu^{2+} 2.8.17
    Fe in $[Fe(CN)_6]^{4-}$ 2.8.18.8   Cu in $[Cu(NH_3)_4]^{2+}$ 2.8.17.8

This assumes that a lone pair of electrons on each CN^- and NH_3 is donated to the central atom. As bonding in complexes is not so straightforward more sophisticated explanations are required.
N.B. Cu^{2+} is sometimes shown as $[Cu(H_2O)_6]^{2+}$ which reacts with aqueous ammonia to become $[Cu(H_2O)_2(NH_3)_4]^{2+}$.

13. pH and pK values

When dealing with very small (or large) numbers it is often convenient to take logarithms.
e.g. $pH = -\log_{10}[H^+(aq)]$ where $[H^+(aq)]$ means concentration in mol l^{-1}.
Similarly dissociation (or hydrolysis) constants K for weak acids and bases can be converted to pK values.
For acetic acid $K = 1 \cdot 8 \times 10^{-5}$
therefore $pK = 5 - \log_{10}(1 \cdot 8) = 4 \cdot 75$
Thus the larger is the pK value, the weaker is the acid.

14. Periodic table

The so-called 'long form' given is probably the most useful arrangement.
i. Elements are divided into 'main' groups (M) and transition elements (T).
ii. In a period containing 'main' elements, the outer electron energy level is filling as the atomic number increases. In a transition metal series an inner level is filling, thus giving rise to variable valency, complex ion formation and coloured salts.
iii. Elements to the left of the 'steps' are metals and those to the right are non-metals. This division is by no means rigid but is none the less useful.
iv. Mendeleeff's original group numbers are retained as they give useful indications concerning valencies (oxidation states).
e.g. Group $2M$ elements have a valency of 2
Group $7M$ elements have valencies of 1 or (less often) 7
Group $3T$ elements have valencies which include 3 (e.g. Sc^{3+}).

15. Radii of atoms and ions

Values quoted for different radii have to be treated with care as 'radius' can have more than one meaning and can vary, depending upon the environment.

A SUMMARY OF THE TERMS USED IN THE QUESTIONS

a. Radii of metal atoms
Metals are considered to be spherical atoms in contact.

e.g. Sodium

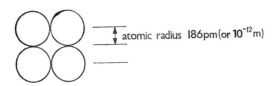

atomic radius 186pm (or 10^{-12}m)

b. Radii of non-metal atoms
i. van der Waals radius
This gives some idea of the size of the atom. The value represents half the distance between the nucleii in two molecules bound by van der Waals forces in a crystal of the solid element.
ii. Covalent radius
This does NOT represent the size of an atom, but these values can be added to give bond lengths.

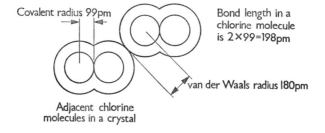

Covalent radius 99pm

Bond length in a chlorine molecule is 2×99=198pm

van der Waals radius 180pm

Adjacent chlorine molecules in a crystal

c. Ionic radii
Ionic radii of metals and non-metals are measured in crystals assuming that the ions are spherical and are touching.

Sodium chloride

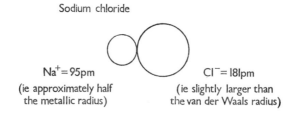

Na^+ = 95pm
(ie approximately half the metallic radius)

Cl^- = 181pm
(ie slightly larger than the van der Waals radius)

It should be noted particularly that the addition of one electron to Cl giving Cl^- does NOT cause the radius to increase from 99 pm (the covalent radius) to 181 pm (the ionic radius) as is sometimes implied in textbooks.

A SUMMARY OF THE TERMS USED IN THE QUESTIONS

16. Organic chemistry
We have tried in our teaching and in these questions to emphasise these points.

a. Application of physical chemistry
Many of the major concepts of physical chemistry are directly applicable to organic reactions.
e.g. Phase equilibria (distillation, eutectics, distribution)
    Laws of equilibria and reaction rate
    Oxidation and reduction
    The mole and equation quantities
Far too often these are learned as 'Physical Chemistry' and never applied elsewhere.

b. Industrial organic chemistry
We have tried to avoid the reactions involved in the industrial production of major chemicals such as ethylene and acetic acid. These are rarely made by a method reproducible in the school laboratory. On the other hand, the properties and uses of the compounds can usually be seen to follow from their molecular structures.

c. Systematic naming of compounds
We strongly urge the use of I.U.P.A.C. nomenclature. The longest carbon-carbon chain length is first found and the name based on that.

chain length	systematic name of hydrocarbon
C	Methane
C—C	Ethane
C—C—C	Propane
C—C—C—C	Butane
C—C—C—C—C	Pentane
C—C—C—C—C—C	Hexane

The position of other groups on this chain is indicated by numbers – always taking the smallest available.

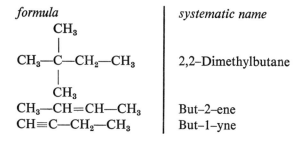

formula	systematic name
$CH_3-\underset{\underset{CH_3}{\mid}}{\overset{\overset{CH_3}{\mid}}{C}}-CH_2-CH_3$	2,2–Dimethylbutane
$CH_3-CH=CH-CH_3$	But–2–ene
$CH\equiv C-CH_2-CH_3$	But–1–yne

A SUMMARY OF THE TERMS USED IN THE QUESTIONS

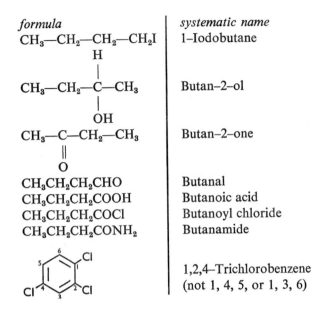

As trivial names are so widespread, however, we have felt it necessary to use a mixture of old and new nomenclature.

d. Functional groups
Most of the properties of alcohols arise from the reactions of the —OH group. It is much more important to know the relatively few reactions of groups such as —OH, —Cl, >C=O, and to be able to apply them in any compound, than to learn the detailed chemistry of a particular compound.

17. Reaction mechanisms
Organic chemistry can easily become a catalogue of facts to be memorised. The facts remain, but a knowledge of how such reactions take place greatly helps the student to learn them. A selection of the most common mechanisms is given below. It is important to note that different conditions can give rise to different mechanisms for the same reaction.

a. Halogenation of alkanes
The chlorination of methane is known to take place by a free radical mechanism. It is likely that substitution reactions between halogens and other alkanes proceed similarly.

A SUMMARY OF THE TERMS USED IN THE QUESTIONS

e.g.
$$:\!\ddot{\underset{..}{Cl}}\!:\!\ddot{\underset{..}{Cl}}\!: \xrightarrow{\text{UV light}} :\!\ddot{\underset{..}{Cl}}\!\cdot + :\!\ddot{\underset{..}{Cl}}\!\cdot$$

$$:\!\ddot{\underset{..}{Cl}}\!\cdot + H\!:\!\underset{H}{\overset{H}{\ddot{C}}}\!:H \longrightarrow H\!:\!\ddot{\underset{..}{Cl}}\!: + H\!:\!\underset{H}{\overset{H}{\ddot{C}}}\!\cdot \quad \overset{\longleftarrow}{\underset{\longleftarrow}{}} \text{Free Radicals}$$

$$H\!:\!\underset{H}{\overset{H}{\ddot{C}}}\!\cdot + :\!\ddot{\underset{..}{Cl}}\!:\!\ddot{\underset{..}{Cl}}\!: \longrightarrow H\!:\!\underset{H}{\overset{H}{\ddot{C}}}\!:Cl + :\!\ddot{\underset{..}{Cl}}\!\cdot$$

b. Addition to alkenes and alkynes
The essential difference between the alkanes and unsaturated hydrocarbons is the high electron density between the carbon atoms in double and treble bonds. It is not surprising that the usual reagents which attack such bonds ($KMnO_4$, halogens, O_3, conc. H_2SO_4) are oxidising agents (i.e. electron acceptors) or are molecules like H—I which may be polarised during the reaction.
e.g. Addition of HI(g) to C_2H_4(g) proceeds thus:

$$\underset{H}{\overset{H}{>}}\!C\!=\!C\!\underset{H}{\overset{H}{<}} \quad \underset{H-I}{\overset{\text{APPROACHING MOLECULES}}{\longrightarrow}} \quad \underset{\underset{\delta+\ \ \delta-}{H\ \ I}}{\overset{H\ \ \delta-\ \ \delta+\ \ H}{>C=C<}} \to \underset{H\ \ I^-}{\overset{H\ \ H}{H-C-\overset{+}{C}-H}} \to \underset{H\ \ H}{\overset{H\ \ H}{H-C-C-H}}$$

c. Hydrolysis of alkyl halides
There are two main mechanisms:
(i) Rate of reaction $\propto$ (Concentration of R — X) × (Concentration of OH^-)
(ii) Rate of reaction $\propto$ (Concentration of R — X) but independent of OH^- concentration.
The hydrolysis of methyl bromide with aqueous alkali follows mechanism (i)
i. $HO^- + CH_3 - Br \to (HO\cdots CH_3\cdots Br)^- \to HO-CH_3 + Br^-$
      Transition state
With more highly branched alkyl halides then mechanism (ii) often applies, as here the carbonium ion formed is stabilised somewhat by electron donation from the methyl groups.

$$\underset{CH_3}{\overset{CH_3}{>}}\!\underset{}{C}\!-\!Br \xrightarrow[\text{STEP}]{\text{RATE DETERMINING}} \underset{CH_3}{\overset{CH_3}{>}}\!C^+ + Br^-$$

A CARBONIUM ION

A SUMMARY OF THE TERMS USED IN THE QUESTIONS

$$CH_3\text{-}C^+(CH_3)(CH_3) + OH^- \xrightarrow{\text{FAST STEP}} CH_3\text{-}C(CH_3)(CH_3)\text{-}OH$$

d. Compounds containing the carbonyl group ($>C=O$)

i. Aldehydes and ketones
Oxygen is more electronegative than carbon so electrons are not shared evenly.
i.e.
$$\overset{\delta+}{C} = \overset{\delta-}{O}$$

This means that in the reactions between this group and the usual reagents (NH_2OH, NH_3, HCN, N_2H_4, etc.), the first stage involves donation of a lone pair of electrons to the carbon atom.
e.g. Cyanohydrin formation:

$$\overset{\delta+}{C}=\overset{\delta-}{O} \rightarrow -\underset{\underset{CN^-}{\uparrow}}{C}=O \rightarrow -\underset{CN}{C}-O^- \xrightarrow{H^+} -\underset{CN}{C}-OH$$

CN^-

ii. Acids and their derivatives
The introduction of an atom such as O, N, or Cl into a carbonyl group means that the electron movement is even more pronounced.

$$-\underset{\text{ELECTRON MOVEMENT}}{C}\overset{O}{\underset{O-H}{\diagdown}} \rightleftharpoons \left(-C\overset{O^-}{\underset{O}{\diagdown}}\right) + H^+_{(aq)}$$

The consequences are that
1. acids give H^+(aq) in solution whereas alcohols do not.
2. the usual properties of the carbonyl group are missing (e.g. no precipitate with 2,4-dinitrophenylhydrazine, etc.)
This also applies to

$$-C\overset{O}{\underset{NH_2}{\diagdown}} \quad \text{amides}$$

A SUMMARY OF THE TERMS USED IN THE QUESTIONS

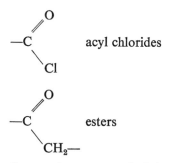

acyl chlorides

esters

None of these gives the reactions expected of the carbonyl group.

e. Amides and amines

Ammonia is a base because the lone pair of electrons on the nitrogen atom can be donated to a proton.

$$H-\overset{H}{\underset{H}{N}}: + H^+ + Cl^- \rightarrow \left(H-\overset{H}{\underset{H}{N}}-H \right)^+ Cl^-$$

Ammonium salt

The same is true of amines (substituted ammonia).

$$\overset{CH_3}{\underset{H}{\underset{H}{N}}}: + H^+ + Cl^- \rightarrow \left(H-\overset{CH_3}{\underset{H}{N}}-H \right)^+ Cl^-$$

Methylammonium chloride

Addition of excess alkali will liberate the free amine from the salt.

With amides, however, the lone pair of electrons is not available in this way and salt formation is therefore rare.

$$CH_3-C\overset{O}{\underset{NH_2}{\diagdown}}$$

ELECTRON MOVEMENT

f. Benzene and other aromatic compounds

There are six electrons spread around the ring of six carbon atoms and this arrangement gives benzene its unusual properties.

A SUMMARY OF THE TERMS USED IN THE QUESTIONS

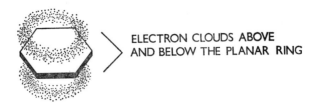

ELECTRON CLOUDS ABOVE
AND BELOW THE PLANAR RING

Some of the consequences of this are as follows:
i. Addition reactions are rare (as there are no 'fixed' double bonds).
ii. Substitution reactions are the most common. Because of the high electron density in the ring, the attacking reagent is usually a positive ion (e.g. NO_2^+ in nitration to give nitrobenzene $C_6H_5NO_2$).
iii. Reactions of some of the common functional groups are modified when attached directly to the ring, especially —Cl, —Br, —I, —NH_2, —OH.
e.g. Phenol (carbolic acid) ionises slightly in solution.

$$\text{C}_6\text{H}_5\text{O-H} \rightleftharpoons \text{C}_6\text{H}_5\text{O}^- + H^+_{(aq)}$$

There is no such tendency in methanol.
iv. Representation of benzene.
I.U.P.A.C. recommends that the benzene ring should be drawn with formal double bonds:

i.e.

We feel that this has disadvantages in teaching and have used instead a version which emphasises the de-localised nature of the electrons and bonds in the molecule.

i.e.

QUESTIONS

Advice to the student on the use of these questions

These questions are designed to help you test yourself on a variety of topics which cover most of the syllabuses for A Level, and are meant to test your understanding rather than your memory.

Each question has from one to five correct answers. When you have finished a set of questions check your answers against the answers and comments at the end of the book. In most cases there will be no argument about the correct answers but in a few questions there may be room for some discussion. We would advise against trying to answer more than one topic at a time, these questions having been proved to be a stimulant in small doses.

The revision questions each have only one correct answer and each series covers a fairly wide field.

ACID – ALKALI TITRATIONS

1. 25 ml of M sodium hydroxide neutralises
   a. 1/40 litre of M hydrochloric acid
   b. 1/40 g of acetic acid
   c. 1/40 mol of acetic acid
   d. 0·025 mol of sulphuric acid
   e. 5 ml of 5 M nitric acid

2. 25 ml of M hydrochloric acid reacts completely with
   a. 50 ml of 0·5 M sodium carbonate
   b. 1 litre of 0·05 M barium hydroxide
   c. 1/40 mol of calcium oxide
   d. 1/40 mol of ammonia
   e. 0·025 mol of lithium hydroxide

3. 1 litre of M amido sulphuric acid reacts completely with
   a. 40 g of sodium hydroxide
   b. 1 mol of sodium carbonate
   c. 1 g-equivalent of potassium carbonate
   d. 106 g of sodium carbonate
   e. 1 litre of M sodium carbonate

4. 1 litre of 0·1 M sodium hydrogen sulphate ($NaHSO_4$) reacts completely with
   a. 4·0 g of sodium hydroxide
   b. 0·1 mol of barium hydroxide
   c. 1·7 g of ammonia
   d. 50 ml of M potassium carbonate
   e. 50 ml of 2 M sodium carbonate

5. 5 g of calcium carbonate reacts completely with
   a. 50 ml M HCl
   b. 100 ml 0·5 M H_2SO_4
   c. 150 ml M H_3PO_4
   d. 100 ml 0·5 M HCl
   e. 200 ml 0·5 M H_2SO_4

6. 1·15 g of sodium were added to water and after the reaction had ceased the solution was made up to 250 ml. 25 ml of this solution were neutralised by
   a. 5·0 ml of M hydrochloric acid
   b. 2·5 ml of M sulphuric acid
   c. 25 ml of 0·4 M hydrochloric acid
   d. 1/200 mol of HCl
   e. 0·005 mol of H_2SO_4

7. 2·0 g of $CaCO_3$ were dissolved in 50 ml of M hydrochloric acid and made up to 250 ml with water. 25 ml of this solution reacted completely with
   a. 40 ml 0·1 M NaOH
   b. 20 ml 0·1 M KOH
   c. 5 ml 0·1 M $Ba(OH)_2$
   d. 5 ml 0·1 M NH_4OH
   e. 1 ml M LiOH

ACID – ALKALI TITRATIONS

8. 4·9 g of potassium hydroxide were made up to 1 litre and titrated against a strong monobasic acid so that 25 ml of acid neutralised 20 ml of alkali.
   Which statements are correct?

   a. Potassium hydroxide is 7/8 M.
   b. Potassium hydroxide is 0·0875 M.
   c. The acid solution is more concentrated than the alkali solution.
   d. The acid is 0·07 M.
   e. The acid is 7/64 M.

9. When titrating sodium hydroxide against propionic acid which of the following indicators are suitable?

   a. Phenolphthalein
   b. Methyl orange
   c. Screened methyl orange
   d. Methyl red
   e. Litmus

10. When titrating sodium carbonate against hydrochloric acid which of the following indicators are suitable to show complete conversion of sodium carbonate to sodium chloride?

    a. Phenolphthalein
    b. Methyl orange
    c. Litmus
    d. Phenol red
    e. Bromothymol blue

11. 1·89 g of a monobasic acid were made up to 250 ml with water and 25 ml of this solution neutralised 20 ml 0·1 M NaOH. The acid could have been

    a. Propanoic acid
    b. Butanoic acid
    c. Stearic acid
    d. Hydrochloric acid
    e. Monochloroacetic acid

12. 0·130 g of a dibasic organic acid formed a silver salt which on ignition produced 0·270 g of silver. The acid could have been

    a. Adipic acid
    b. Benzoic acid
    c. Oxalic acid
    d. Malonic acid
    e. Succinic acid

13. 2·44 g of a monobasic organic acid were dissolved in 50 ml M NaOH. The solution was made up to 250 ml and 25 ml of this diluted solution required 30 ml 0·1 M HCl to neutralise the excess NaOH. The acid could have been

    a. Propanoic acid
    b. Dichloroacetic acid
    c. Bromoacetic acid
    d. Benzoic acid
    e. Terephthalic acid

14. An organic acid forms a silver salt, 0·362 g of which on ignition produce 0·216 g of silver. It also gives an ester which has a vapour density of about 50. The most likely ester is:

    a. Methyl propanoate
    b. Ethyl acetate
    c. Ethyl propanoate
    d. Propyl acetate
    e. Butyl formate

ACID − ALKALI TITRATIONS

15. 34 ml of 0·5 M HCl were required to neutralise 10 ml of household ammonia. The ammonia solution contains
    a. 1·7 mol l^{-1} of NH_3.
    b. 1·7 mol of NH_3.
    c. 38 litres of NH_3 per litre (at s.t.p.).
    d. 17 g of NH_3 per litre.
    e. 1·7% w/w of NH_3.

16. 2·5 g of a mixture of NaCl and NaOH were made up to 250 ml with water. 25 ml of this solution neutralised 20 ml 0·1 M HCl. The mixture contained
    a. 1·25 g of sodium chloride.
    b. 50% by weight of sodium chloride.
    c. 0·02 mol NaOH.
    d. 0·80 g of NaOH.
    e. 1/40 mole of NaOH.

17. 1·7 g of a mixture of KOH and NaOH were made up to 250 ml with water. 25 ml of this solution neutralised 37·5 ml 0·1 M HCl. The mixture contained
    a. 37·5 × 10^{-3} mol OH$^-$ ions in 250 ml of solution.
    b. twice the weight of KOH as NaOH.
    c. 0·15 mol l^{-1} of alkali.
    d. 1·0 g of KOH.
    e. 0·7 g of KOH.

18. A solution of NaOH and Na_2CO_3 was titrated with 0·1 M HCl. 25 ml of the solution required 16 ml 0·1 M HCl with phenolphthalein as indicator, but required 24 ml 0·1 M HCl with methyl orange as indicator. The mixture contained:
    a. 0·32 mol NaOH in 1 litre.
    b. moles of Na_2CO_3: moles of NaOH = 2:1.
    c. g of Na_2CO_3: g of NaOH = 53:40.
    d. 0·32 mol of Na_2CO_3 per litre.
    e. 12½ g of NaOH per litre.

19. 2·8 g of animal feedstuff when heated with conc. H_2SO_4 produced ammonium sulphate. This on boiling with NaOH evolved ammonia which neutralised 7·5 ml M H_2SO_4. There were therefore:
    a. 0·15 mol of NH_3 evolved.
    b. 0·255 g of NH_3 evolved.
    c. 0·210 g of nitrogen in the sample.
    d. 7½% w/w of nitrogen in the sample.
    e. 15% w/w of nitrogen in the sample.

20. 0·1 mole of amyl acetate was mixed with dilute H_2SO_4 so that the total volume of liquid was 250 ml. 25 ml of this solution initially neutralised 2 ml 0·1 M NaOH. After refluxing for a short time 25 ml were equivalent to 18 ml 0·1 M NaOH. Which statements are correct?
    a. The total acid content before refluxing was 0·02 mol H$^+$ ions.
    b. 18 ml M acetic acid were produced.
    c. 0·016 mol H$^+$ ions were formed.
    d. 0·016 mol of acetic acid was produced.
    e. 16% by weight of the ester was hydrolysed.

OXIDATION NUMBERS AND REDOX TITRATIONS

1. Which of the following are oxidations?
   a. $MnO_4^- \rightarrow Mn^{2+}$
   b. $MnO_4^- \rightarrow MnO_2$
   c. $MnO_4^- \rightarrow MnO_4^{2-}$
   d. $MnO_2 \rightarrow MnO_4^-$
   e. $MnO_2 \rightarrow Mn^{2+}$

2. Which of the following are reductions?
   a. $Zn \rightarrow Zn^{2+}$
   b. $Fe^{2+} \rightarrow Fe^{3+}$
   c. $H_2S \rightarrow S$
   d. $Na^+ \rightarrow Na$
   e. $V_2O_5 \rightarrow NH_4VO_3$

3. Which of the following are oxidations?
   a. $H_2O_2 \rightarrow 2OH^-$
   b. $CrO_4^{2-} \rightarrow Cr_2O_7^{2-}$
   c. $[Fe(CN)_6]^{3-} \rightarrow [Fe(CN)_6]^{4-}$
   d. $IO_3^- \rightarrow I^-$
   e. $2H^- \rightarrow H_2$

4. Which of the following elements are oxidised?
   a. Potassium in water.
   b. Magnesium in dilute sulphuric acid.
   c. Copper in silver nitrate solution.
   d. Aluminium in sodium hydroxide solution.
   e. Copper anode in the electrolysis of copper sulphate solution.

5. Which of the following are reductions?
   a. $SO_3^{3-} \rightarrow SO_4^{2-}$
   b. $S^{2-} \rightarrow S$
   c. $2S_2O_3^{2-} \rightarrow S_4O_6^{2-}$
   d. $S_2O_3^{2-} \rightarrow 2S$
   e. $S_2O_3^{2-} \rightarrow 2SO_4^{2-}$

6. In which of the following compounds is the oxidation number of sulphur $+6$?
   a. $NaHSO_3$.
   b. Na_2SO_4.
   c. Na_2SO_3.
   d. $Na_2S_4O_6$.
   e. $Na_2S_2O_8$.

7. Which statements are correct concerning the compound $K_3Mo(CN)_8$
   a. Oxidation number of Mo is $+3$.
   b. Oxidation number of Mo is $+5$.
   c. Oxidation number of Mo is $+8$.
   d. Co-ordination number of Mo is 3.
   e. Co-ordination number of Mo is 8.

OXIDATION NUMBERS AND REDOX TITRATIONS

8. When sodium nitrite is added to acid permanganate the half equation is: (Electrons have been omitted)
   a. $NO_2^- \rightarrow NO_2$
   b. $NO_2^- + 2H^+ \rightarrow NO + H_2O$
   c. $NO_2^- + H_2O \rightarrow NO_3^- + 2H^+$
   d. $2NO_2^- + 6H^+ \rightarrow N_2O + 3H_2O$
   e. $2NO_2^- + 8H^+ \rightarrow N_2 + 4H_2O$

9. In which cases would 1 litre of solution oxidise 1 mol of Fe^{2+} ions to Fe^{3+} ions?
   a. M/2 hydrogen peroxide in excess sulphuric acid.
   b. M/6 potassium permanganate in excess sulphuric acid.
   c. M/6 potassium dichromate in excess sulphuric acid.
   d. M iodine in excess potassium iodide.
   e. M/5 neutral potassium permanganate.

10. Which of the following would oxidise 0·56 g of Fe^{2+} ions to Fe^{3+} ions in the presence of excess sulphuric acid?
    a. 0·002 mol permanganate solution.
    b. 0·004 mol dichromate solution.
    c. 0·01 mol permanganate solution.
    d. 10 ml 0·2 M permanganate solution.
    e. 10 ml 0·2 M dichromate solution.

11. When hydrogen peroxide is oxidised by potassium permanganate, which half equations represent the change which takes place?
    a. $2e^- + H_2O_2 \rightarrow 2OH^-$
    b. $2e^- + 2H^+ + H_2O_2 \rightarrow 2H_2O$
    c. $2H_2O_2 \rightarrow 2H_2O + O_2$
    d. $H_2O_2 \rightarrow 2H^+ + O_2 + 2e^-$
    e. $2OH^- + H_2O_2 \rightarrow 2H_2O + O_2 + 2e^-$

12. When sodium thiosulphate is oxidised by iodine then
    a. sodium tetrathionate is formed.
    b. the reaction is represented by
       $2S_2O_3^{2-} \rightarrow S_4O_6^{2-} + 2e^-$
    c. 1 mol $Na_2S_2O_3$ releases 2 mol of electrons.
    d. the same change takes place if we oxidise with chlorine.
    e. acid is always placed in the thiosulphate first.

13. A solution composed of 0·68 g of hydrogen peroxide in 100 ml of aqueous solution
    a. contains 0·2 mol H_2O_2.
    b. is an 0·2 M solution.

c. decolourises 400 ml of 0·02 M acid permanganate.
d. decolourises 80 ml of 0·01 M acid permanganate.
e. is a 4·48 volume solution.

14. Which of the following are redox reactions?

   a. Silver nitrate solution is added to sodium chloride solution.
   b. Potassium chromate solution is added to lead nitrate solution.
   c. Potassium dichromate solution is added to sodium hydroxide solution.
   d. Copper oxide is added to dilute sulphuric acid.
   e. Zinc is placed in lead acetate solution.

15. Which of the following reacts completely with 25 ml of 0·1 M oxalic acid?

   a. 25 ml 0·1 M acid $KMnO_4$.
   b. 10 ml 0·1 M acid $KMnO_4$.
   c. 5 ml 0·1 M acid $KMnO_4$.
   d. 25 ml 0·1 M NaOH.
   e. 25 ml 0·1 M $Ba(OH)_2$.

16. Which of the following solutions will completely oxidise 25 ml of an acid solution of 0·1 M iron(II) oxalate?

   a. 25 ml 0·1 M $KMnO_4$.
   b. 25 ml 0·2 M $KMnO_4$.
   c. 25 ml 0·6 M $KMnO_4$.
   d. 15 ml 0·1 M $KMnO_4$.
   e. 25 ml 0·06 M $KMnO_4$.

17. 25 ml of a saturated solution of SO_2 in water required 300 ml of 0·1 M $KMnO_4$ to oxidise it to SO_4^{2-} under laboratory conditions. Assuming that 1 mol of SO_2 gas occupies 24 litre under laboratory conditions which of the following statements are true?

   a. $SO_2 + 2H_2O \rightarrow SO_4^{2-} + 4H^+ + 2e^-$.
   b. 5 mol SO_2 reduces 2 mol $KMnO_4$.
   c. 1 litre of saturated SO_2 solution reduces 1·2 mol $KMnO_4$.
   d. 1 litre saturated SO_2 contains 4·8 mol SO_2.
   e. 1 litre of water dissolves 72 litre of SO_2 under laboratory conditions.

18. 0·02 mol of acetone $(CH_3)_2CO$ is added to 250 ml of 0·1 M I_2 in the presence of dilute acid and left for several days. At the end of this time 100 ml of 0·1 M $Na_2S_2O_3$ are required to decolourise the solution.
Which of the following statements are true?

   a. 1 mol I_2 oxidises 1 mol $Na_2S_2O_3$.
   b. 100 ml 0·1 M $Na_2S_2O_3$ reduces 50 ml 0·1 M I_2.
   c. 0·02 mol acetone reacts with 0·02 mol I_2.
   d. 1 mol acetone reacts with 1 mol I_2.
   e. 1 atom of iodine is substituted into 1 molecule of acetone.

19. A blue compound
   $Na_3MnO_4, 7H_2O$
   is obtained by the reaction of sodium manganate with sodium formate in alkaline solution.
   Which statements are correct?

   a. The manganate ion is reduced.
   b. The formate ion is reduced.
   c. The oxidation state of manganese in the blue compound is $+5$.
   d. The valency state of manganese has increased from $+2$ to $+3$.
   e. The compound is sodium tetraoxomanganate(V) heptahydrate.

20. An 0·1 M solution of ammonium metavanadate (NH_4VO_3) was reduced with acid and a metal amalgam. 25 ml of this solution required 15 ml of 0·1 M potassium permanganate to oxidise it (in acid solution) back to its original form. Hence:

   a. the reduced form was vanadium(II).
   b. the reduced form was vanadium(III).
   c. the reduced form was vanadium(IV).
   d. the oxidation number of vanadium was decreased by 3 in the reduction.
   e. the oxidation number of vanadium was decreased by 2 in the reduction.

MOLECULES, MOLES, AND EQUILIBRIA

The following information may be useful

Avogadro constant (L).........6×10^{23} mol^{-1}
Volume of one mole of gas at s.t.p..........22·4 dm^3 mol^{-1}
Standard temperature and pressure (s.t.p.).........10^5 Nm^{-2} (or 760mmHg) and 273K.

1. Calculate the number of molecules in 1 cm^3 of helium gas at s.t.p.
   a. $1·34 \times 10^{25}$
   b. $1·34 \times 10^{22}$
   c. $2·68 \times 10^{22}$
   d. $3·73 \times 10^{20}$
   e. $2·68 \times 10^{19}$

2. One cm^3 of a vegetable oil of molecular weight 480 and specific gravity 0·8 is dropped onto water and spreads out in a film one molecule thick. Which statements are correct?
   a. There is $\frac{1}{600}$ mole of oil on the water.
   b. There are 10^{21} molecules of oil in 1 cm^3.
   c. The thickness of the film is 10^{-9}m.
   d. The area of the film is 100 square metres.
   e. The oil could be glycerine.

3. 0·3 g of carbon is burned in oxygen. Which statements are correct?
   a. $\frac{1}{40}$ mole of CO_2 is produced.
   b. $1·5 \times 10^{22}$ molecules of CO_2 are formed.
   c. $5·6 \times 10^{25}$ cm^3 of CO_2 at s.t.p. are formed.
   d. $\frac{1}{40}$ mole of O_2 is used.
   e. 0·4 g of oxygen is used.

4. Compressed oxygen is sold in cylinders of 40 dm^3 capacity and 130 atmospheres pressure at 0°C. How many moles of oxygen does a cylinder contain?
   a. 232
   b. 5200
   c. 1 164 800
   d. $1·39 \times 10^{26}$
   e. $4·368 \times 10^{25}$

5. An evacuated bulb weighs 108·16 g. Filled with oxygen at s.t.p. it weighs 109·76 g, but when filled with the gas from a volcano at s.t.p. it weighs 111·36 g. Indicate the gas(es) which the sample could be.
   a. SO_2
   b. SiF_4
   c. Equal volumes of Si_2H_6 and H_2S
   d. S_8
   e. NF_3

6. What is the maximum decrease in volume when 10 cm³ of nitrogen are sparked with 100 cm³ of hydrogen? (Temperature and pressure remain constant at 273K and $10^5 Nm^{-2}$.)

   a. 15 cm³
   b. 20 cm³
   c. 25 cm³
   d. 30 cm³
   e. 35 cm³

7. The following gases are contained in vessels of 2 dm³ capacity. Which have concentrations of 0·05 mol dm⁻³?

   a. 0·1 g of hydrogen
   b. 1·4 g of nitrogen
   c. 0·8 g of oxygen
   d. 0·4 g of helium
   e. 3·55 g of chlorine

8. It is often stated that 20% of the air is oxygen. This is better stated as:

   a. 20% of the weight is oxygen.
   b. 20% of the volume is oxygen.
   c. the partial pressure of oxygen is 0·2 atmosphere.
   d. 20% of the molecules are oxygen molecules.
   e. 80% of the air is nitrogen.

9. A closed vessel contains equal numbers of oxygen and hydrogen molecules. Which of the following is true?

   a. The average speed of the hydrogen molecules is greater.
   b. The hydrogen molecules strike the sides more often.
   c. The average kinetic energies of the two gases is the same.
   d. Equal numbers of moles of each are present.
   e. The partial pressures of the gases are the same.

10. A gaseous compound of molar mass 210 dissociates on heating into two gaseous compounds so that the vapour density is 70. What is the extent of dissociation if we assume that $AB \rightleftharpoons A + B$?

    a. 30%
    b. 33·3%
    c. 50%
    d. 60%
    e. 66·7%

11. The molar mass of acetic acid when dissolved in benzene is found to be about 120. This could be due to:

    a. dissociation of acetic acid into ions.
    b. association of acetic acid into double molecules.
    c. formation of a carbon isotope.
    d. formation of an isomer of acetic acid.
    e. partial association of acetic acid into triple molecules.

MOLECULES, MOLES, AND EQUILIBRIA

12. The following masses of gases are all contained in the same closed iron vessel so that the total pressure is $10^5 Nm^{-2}$. Which gases have partial pressures of $2.5 \times 10^4 Nm^{-2}$?

    a. 0·2 g of hydrogen
    b. 1·4 g of nitrogen
    c. 3·2 g of oxygen
    d. 0·4 g of helium
    e. 3·55 g of chlorine

13. Which of the following represent closed systems in dynamic equilibrium?

    a. a saturated solution of lime water.
    b. a 0·1 molar solution of acetic acid.
    c. a solution of the complex salt $Cu(NH_3)_4SO_4$ in water.
    d. a molar solution of sodium chloride.
    e. a few drops of water at the top of a barometer tube.

14. When performed in closed vessels at 200°C, which of the following reactions would you consider to be systems in equilibrium?

    a. $H_2 + Cl_2 = 2HCl$
    b. $P_4 + 5O_2 = P_4O_{10}$
    c. $F_2 + 2K = 2KF$
    d. $CaCO_3 = CO_2 + CaO$
    e. $NH_4Cl = NH_3 + HCl$

15. Which of the following reactions are favoured in the forward direction by an increase in pressure on the system (temperature remaining constant)?

    a. $2NO_2 \rightleftharpoons N_2O_4$
    b. $H_2 + I_2 \rightleftharpoons 2HI$
    c. $SO_2 + Cl_2 \rightleftharpoons SO_2Cl_2$
    d. $CO + 2H_2 \rightleftharpoons CH_3OH$
    e. $C_2H_4 + H_2 \rightleftharpoons C_2H_6$

16. Given that the reaction
    $2SO_2 + O_2 \rightleftharpoons 2SO_3 : \Delta H = -196$ kJ
    takes place in a closed system, would the percentage of SO_3 in the equilibrium mixture be increased by:

    a. increasing the temperature?
    b. increasing the pressure on the system?
    c. adding a catalyst?
    d. adding more SO_2?
    e. removing oxygen?

17. Given that the reaction
    $N_2 + 3H_2 \rightleftharpoons 2NH_3 : \Delta H = -50$ kJ
    takes place in a closed system, would the percentage of NH_3 in the equilibrium mixture be increased by:

    a. increase of temperature?
    b. increase of pressure?
    c. adding nitrogen?
    d. adding helium?
    e. adding a catalyst?

18. In which of the following reactions do you think a catalyst is necessary?

a. $CO + H_2O \rightleftharpoons CO_2 + H_2$: $\Delta H = -42$ kJ
b. $4HCl + O_2 \rightleftharpoons 2H_2O + 2Cl_2$: $\Delta H = -113$ kJ
c. $N_2 + O_2 \rightleftharpoons 2NO$: $\Delta H = +89$ kJ
d. $2NO_2 \rightleftharpoons 2NO + O_2$: $\Delta H = +113$ kJ
e. $CaC_2 + N_2 \rightleftharpoons CaCN_2 + C$: $\Delta H = -302$ kJ

19. 40 ml of acetic acid and 60 ml of propanol are placed together in a flask and allowed to reach equilibrium. How can the concentration of propyl acetate be increased?

a. Doubling the pressure.
b. Increasing the volume of acid to 80 ml.
c. Increasing the volume of propanol to 120 ml.
d. Increasing the volume of acetic acid to 60 ml and of propanol to 90 ml.
e. Adding HCl gas as a catalyst.

20. Given the following gaseous reactions, which statements are correct?
   i. $C_2H_4 + H_2 \rightleftharpoons C_2H_6$
   ii. $H_2 + I_2 \rightleftharpoons 2HI$
   iii. $2SO_2 + O_2 \rightleftharpoons 2SO_3$
   iv. $N_2 + 3H_2 \rightleftharpoons 2NH_3$

a. K_p in ii. is independent of the total pressure (p).
b. K_p is the same as K_c only for reaction iii.
c. In reactions i. and iii. $K_c \propto \dfrac{1}{volume}$.
d. In reaction iv. $K_p \propto p^2$.
e. In reaction i. $K_p \propto p$.

COLLIGATIVE PROPERTIES

The following information may be useful:

solvent	freezing point	boiling point	ebullioscopic constant K mol^{-1}kg	cryoscopic constant K mol^{-1}kg
Water	0°C	100°C	0·52	1·86
Benzene	5·4°C	80·4°C	2·57	5·12
Acetone	−94°C	57·5°C	1·72	—
Acetic acid	17°C	118°C	3·1	3·9
Camphor	—	—	—	40

1. Which of the following are dependent on the number of solute particles in a given weight of solvent?
   a. Osmotic pressure.
   b. Elevation of boiling point of solvent.
   c. Depression of freezing point of solvent.
   d. Vapour density of solvent.
   e. Vapour pressure of solute.

2. The saturated pressure of vapour from 100 g of water in which urea has been dissolved is directly proportional to the:
   a. molecular weight of urea.
   b. number of moles of urea.
   c. mole fraction of solvent.
   d. absolute temperature.
   e. boiling point of urea.

3. The rules relating to vapour pressure which enable us to find the molecular weight of a solute are restricted to (or by):
   a. negligible vapour pressure of the solvent.
   b. non-volatile solute.
   c. dilute solution.
   d. constant temperature.
   e. no change in the molecular form of the solute when it dissolves in the solvent.

4. Calculate the vapour pressure at 22°C of an aqueous solution containing 0·2 mole of sugar in 1 mole of solution if the vapour pressure of pure water is 2·60 kN m^{-2} at 22°C.
   a. 0·52 kN m^{-2}
   b. 1·50 kN m^{-2}
   c. 2·08 kN m^{-2}
   d. 2·72 kN m^{-2}
   e. 3·20 kN m^{-2}

5. Calculate the vapour pressure at 20°C of an aqueous solution of glycol $(CH_2OH)_2$ containing 62 g in 152 g of solution. (Vapour pressure of pure water at 20°C is 2·4 kN m^{-2}.)
   a. 1·8 kN m^{-2}
   b. 2·0 kN m^{-2}
   c. 2·2 kN m^{-2}
   d. 2·8 kN m^{-2}
   e. 3·0 kN m^{-2}

COLLIGATIVE PROPERTIES

6. Which of the following produces the greatest lowering of the freezing point of 1 kg of water?
   a. 100 g of sucrose, $C_{12}H_{22}O_{11}$
   b. 60 g of glycerol, $C_3H_8O_3$
   c. 80 g of glucose, $C_6H_{12}O_6$
   d. 50 g of glycol, $C_2H_6O_2$
   e. 0·5 mole of fructose, $C_6H_{12}O_6$

7. 1 g of urea (CON_2H_4) is dissolved in 30 g of water at 20°C. Which statements are correct?
   a. Mole fraction of solute is 1/31.
   b. Mole fraction of solvent is 0·99.
   c. Solution contains 1 mole of urea in 1·8 litres of water.
   d. Osmotic pressure is about 12 atmospheres.
   e. Solution is 1/18 M.

8. A given mass of an organic compound dissolved in 100 g of benzene lowered the freezing point of benzene by 1·28°C. If the same mass was dissolved in 100 g of acetic acid, 100 g of acetone or 100 g of camphor then:
   a. acetone solution would boil at 61·8°C.
   b. freezing point of camphor would be lowered by 10°C.
   c. acetic acid solution would boil at 117°C.
   d. acetic acid solution would freeze at 6·5°C.
   e. 1/4 mole of solute was dissolved in each case.

9. A theoretical value can be obtained for the molar mass of a compound by calculation from the chemical formula. Which of the following will give practical results which agree within 5 per cent of the calculated molar mass when the freezing point of an aqueous solution is determined?
   a. Sodium chloride
   b. Sodium acetate
   c. Acetic acid
   d. Acetamide
   e. Acetyl chloride

10. A solution of acetamide in 100 g of water begins to freeze when cooled to −0·24°C. If the temperature drops to −0·96°C then
    a. pure ice separates out.
    b. the acetamide solution freezes.
    c. the acetamide solution becomes more concentrated.
    d. the acetamide solution becomes three times as concentrated.
    e. 75·0 g of ice are formed.

COLLIGATIVE PROPERTIES

11. 3 g of a pentose ($C_5H_{10}O_5$) are dissolved in 100 g of water. Which statements are correct?

    a. Mole fraction of the solute is $\frac{3}{100}$.
    b. Solution starts to freeze at $-0\cdot104°C$.
    c. Solution starts to boil at $103\cdot7°C$.
    d. Osmotic pressure of about $4\cdot5$ atmospheres at $20°C$ can be produced.
    e. Solution is $0\cdot02$ M.

12. The boiling point of a dilute solution of benzoic acid in benzene is given as $82°C$. This is the value registered by a thermometer:

    a. in the liquid when boiling starts.
    b. in the vapour when boiling starts.
    c. in the vapour when a constant value is reached.
    d. An average value of the thermometer readings in the liquid during distillation.
    e. An average value of the thermometer readings in the vapour during distillation.

13. The freezing point of a dilute solution of acetamide in glacial acetic acid is quoted as $15°C$. This is the value when:

    a. the solution freezes completely.
    b. crystals of acetamide first appear.
    c. crystals of acetic acid first appear.
    d. crystals of acetic acid and acetamide first form together.
    e. crystals of ice appear.

14. An aqueous solution of a non-electrolyte gave an osmotic pressure of 14 atmospheres at $0°C$. Which statements are correct?

    a. There was 1 mole in $1\cdot4$ kg of water.
    b. There was 1 mole in $1\cdot6$ litres of solution.
    c. There was $0\cdot6$ mole in 100 g of water.
    d. Ice would start to separate at $-1\cdot1°C$.
    e. The solution would start to boil at $100\cdot3°C$.

15. $6\cdot0$ g of urea and $9\cdot2$ g of glycerol were dissolved in 100 g of water. The solution would start to freeze at

    a. $-0\cdot186°C$
    b. $-0\cdot372°C$
    c. $-1\cdot86°C$
    d. $-3\cdot72°C$
    e. $-18\cdot6°C$

16. An organic compound is said in the literature to have a molar mass (M_r) of 330 ± 10. Indicate the methods which would give this accuracy using 5·00 g of compound in 100 g of solvent.

a. Depression of the freezing point of water with a thermometer reading to 0·01°C.
b. Elevation of the boiling point of water using a Beckmann thermometer.
c. Osmotic pressure of an aqueous solution measured to the nearest 10 mmHg.
d. Lowering of the saturated vapour pressure of water at 15°C measured to the nearest 0·1 mmHg (s.v.p. of water at 15°C is 13 mmHg.)
e. Using the Rast method with a thermometer reading to 0·1°C (the solvent being camphor).

17. 2·00 g of cinnamic acid ($C_6H_5CH:CHCOOH$) dissolved in 100 g of acetic acid depressed the freezing point by 0·39°C. This corresponds to:

a. expected molar mass.
b. 27 per cent dissociation into ions.
c. 54 per cent dissociation into ions.
d. 52 per cent association into double molecules.
e. 39 per cent association into triple molecules.

18. 7·2 g of cadmium iodide dissolved in 100 g of water depressed the freezing point by 0·36°C. Which is the most probable formula for the compound?

a. CdI_2
b. $Cd^{2+} 2I^-$
c. $(CdI_2)_2$
d. $Cd^{2+}(CdI_4)^{2-}$
e. $(CdI)^+ I^-$

19. 2·3 g of a complex platinum ammine depress the freezing point of 100 g of water by 0·42°C, measured with an accuracy of ±0·05°C. The compound could be

a. $Pt(NH_3)_2Cl_4$
b. $[Pt(NH_3)_5Cl]^{3+} 3Cl^-$
c. $[Pt(NH_3)_4Cl_2]^{2+} 2Cl^-$
d. $[Pt(NH_3)_3Cl_3]^+ Cl^-$
e. $2K^+[PtCl_6]^{2-}$

20. Which of the following would give the greatest lowering of the freezing point of 100 g of water?
(Take 1 g of anhydrous compound in each case.)

a. Sodium chloride
b. Potassium chloride
c. Zinc chloride
d. Urea
e. Calcium chloride

ACIDS AND ACIDITY

You may need some of the following information: (all at 25°C)
  Avogadro constant is 6×10^{23} mol^{-1}
  pK_w for water is 14
  pK_a for acetic acid is 4·8
  pK_a for formic acid is 3·8
  pK_a for propanoic acid is 4·9
  pK_a for monochloroacetic acid is 2·9
  α is the degree of ionisation (dissociation)
You will need logarithm tables.

1. If an M solution of a weak monobasic acid is diluted, then
   a. more molecules dissociate.
   b. more ions are formed.
   c. the degree of ionisation increases.
   d. the molar conductance increases.
   e. the dissociation into ions is 100 per cent at 10 000 litres dilution.

2. If an M solution of a strong monobasic acid is diluted, then
   a. the apparent degree of ionisation increases with dilution.
   b. α appears to be 1 at 10 000 litres dilution.
   c. the molar conductance is directly proportional to the dilution.
   d. the molar conductance reaches a maximum at 10 000 litres dilution.
   e. we can say that α is approximately 1 in quick calculations for concentrations below 0·1 M.

3. Ostwald's Dilution Law is obeyed by aqueous solutions of
   a. acetic acid
   b. hydrochloric acid
   c. sodium chloride
   d. methylamine hydrochloride
   e. silver chloride

4. A litre of 0·01 M calcium chloride contains
   a. 10^{-2} mol of calcium chloride
   b. 10^{-2} mol of Ca^{2+} ions
   c. 10^{-2} mol of Cl$^-$ ions
   d. 6×10^{21} Ca^{2+} ions
   e. 6×10^{21} calcium chloride molecules

ACIDS AND ACIDITY

5. Given 0·01 M HCl and 0·1 M acetic acid, then
    a. the acetic acid is the stronger.
    b. the acetic acid is the more concentrated.
    c. the acetic acid solution dissolves magnesium faster.
    d. the pH of the acetic acid is the greater.
    e. the H⁺ concentration in the acetic acid is the greater.

6. The relative strengths of weak acids can be seen by a comparison of their
    a. pH values
    b. H⁺ concentrations
    c. concentrations
    d. dissociation constants
    e. percentages of dissociation

7. For which of the following would you expect to find dissociation constants in a comprehensive book of physical and chemical constants?
    a. Sodium chloride
    b. Sodium acetate
    c. Silver chloride
    d. Perchloric acid
    e. Hypochlorous acid

8. An aqueous solution has a hydrogen ion concentration of 10^{-8} mol l⁻¹, therefore
    a. the solution is weakly acidic.
    b. the pH is 8.
    c. the solution is weakly alkaline.
    d. the pOH is 6.
    e. the hydroxide ion concentration is 10^{-4} mol l⁻¹.

9. Given that pK_w for water at 25°C is 14 and that the reaction $H^+ + OH^- \rightarrow H_2O$ is exothermic, which of these values are of the right order?
    a. K_w is 10^{-14} at 25°C
    b. K_w is 3×10^{-15} at 10°C
    c. pK_w is 14·5 at 10°C
    d. pH of water at 10°C is 7·26
    e. pOH of water at 10°C is 6·94

10. Given 0·01 M acetic acid, then
    a. the concentration is 10^{-2} mol l⁻¹.
    b. the dilution is 10^{-2} litres.
    c. K_a is $10^{-4 \cdot 8}$.
    d. pH is 3·4.
    e. the hydrogen ion concentration is $10^{-2 \cdot 4}$ mol l⁻¹.

ACIDS AND ACIDITY

11. Given 0·01 M propanoic acid, then
    a. K_a is $10^{-4.9}$.
    b. K_a is $1·26 \times 10^{-5}$.
    c. pH is 3·45.
    d. α is less than 5 per cent.
    e. propanoic acid is stronger than acetic acid.

12. Given 0·001 M formic acid, then
    a. K_a is $1·58 \times 10^{-4}$.
    b. pH is 3·4.
    c. α is small, less than 5 per cent.
    d. α is large, greater than 20 per cent.
    e. formic acid is weaker than acetic acid.

13. Given 0·01 M propanoic acid and 0·001 M formic acid, then
    a. formic acid is the stronger.
    b. formic acid is the least concentrated.
    c. the pH is about the same.
    d. there are more molecules per litre of propanoic acid.
    e. the H^+ ion concentration in formic acid is the smaller.

14. One ml of M hydrochloric acid was made up to 10 litres of solution with distilled water, therefore
    a. the dilution was 10 litres.
    b. the concentration was 10^{-1} mol l^{-1}.
    c. the dilution was 10^{-4} l mol^{-1}.
    d. the hydrogen ion concentration was 4.
    e. the pOH was 10.

15. 0·4 g of sodium hydroxide was made up to 100 litres of solution with distilled water, therefore
    a. the concentration was 0·0004 g l^{-1}.
    b. the concentration was 10^{-4} mol l^{-1}.
    c. the dilution was 10^{-4} l mol^{-1}.
    d. the OH^- ion concentration was 4.
    e. the pH was 10.

16. It is found that twenty drops from a burette are the same as one ml. If one drop of 0·2 M HCl is allowed to fall into 100 ml of 0·001 M NaOH, then
    a. the drop has a volume of 5×10^{-5} litre.
    b. the drop contains 10^{-5} mol of hydrochloric acid.
    c. the sodium hydroxide solution contains initially 10^{-4} mol of NaOH.
    d. the pH of the NaOH solution is initially 10.
    e. the pH of the final solution will be 5.

ACIDS AND ACIDITY

17. What is the approximate pH of a solution made by adding 50 ml of 2 M acetic acid to 10 ml of M sodium acetate?

a. 4
b. 5
c. 6
d. 7
e. 8

18. Which of the following solutions would you expect to give a red colouration with litmus solution?

a. 2 M Na_2CO_3
b. M $NaHCO_3$
c. 0·1 M $NaHSO_3$
d. 0·1 M $Al_2(SO_4)_3$
e. 0·1 M Na_2HPO_4.

19. An indicator has a dissociation constant of 10^{-5} and is a weak base. It is green in alkaline solution and red in acid solution. Which deductions are likely to be correct?

a. It would change colour at pH 5.
b. It would be green in pure water.
c. It would be green in M sodium carbonate.
d. It would be green in M ammonium sulphate.
e. It would be green in M ammonium chloride.

20. M hydrochloric acid is added slowly to 50 ml of M sodium hydroxide. The pH of the solution

a. is initially 14.
b. after 49 ml of HCl is added becomes 12.
c. after 49·9 ml of HCl is added becomes 10.
d. after 50·1 ml of HCl is added becomes 3.
e. after 51 ml of HCl is added becomes 2.

IONIC EQUILIBRIUM — SOLUBILITY PRODUCT — STABILITY CONSTANT

1. In a saturated solution of sodium chloride there exists an equilibrium between
   a. solid NaCl and molecules of NaCl in solution.
   b. solid NaCl and hydrated molecules of NaCl.
   c. solid NaCl and hydrated Na^+ and Cl^- ions.
   d. molecules of NaCl in solution and Na^+ and Cl^- ions.
   e. molecules of NaCl in solution and hydrated Na^+ and Cl^- ions.

2. Reference books quote the solubility products of
   a. any insoluble compound.
   b. any compound which dissolves slightly in water.
   c. sparingly soluble salts.
   d. sparingly soluble hydroxides.
   e. partially ionised sparingly soluble compounds.

3. Which of the following would you expect to find in a book of reference containing details of solubility products?
   a. Sodium sulphide
   b. Hydrogen sulphide
   c. Antimony sulphide
   d. Acetic acid
   e. Lead acetate

4. Which equation do you think best illustrates the equilibrium which exists in a saturated solution of barium sulphate?
   a. $BaSO_4 \rightleftharpoons Ba^{2+} + SO_4^{2-}$
   b. $Ba^{2+}SO_4^{2-} \rightleftharpoons Ba^{2+} + SO_4^{2-}$
   c. $BaSO_4(s) \rightleftharpoons BaSO_4(aq)$
   d. $BaSO_4(s) \rightleftharpoons Ba^{2+}(aq) + SO_4^{2-}(aq)$
   e. $BaSO_4(s) \rightleftharpoons Ba^{2+}(l) + SO_4^{2-}(l)$

5. When excess KCN solution is added to $AgNO_3$ solution an equilibrium exists between the silver ions, the cyanide ions and the complex anions which are formed. Which is the most satisfactory way of expressing this equilibrium?
   a. $Ag^+ + 2CN^- \rightarrow Ag(CN)_2^-$
   b. $Ag^+ + 2CN^- \rightleftharpoons Ag(CN)_2^-$
   c. $Ag^+(aq) + 2CN^-(aq) \rightarrow Ag(CN)_2^-(aq)$
   d. $Ag(H_2O)_2^+ + 2CN^- \rightleftharpoons Ag(CN)_2^- + 2H_2O$
   e. $2KCN + AgNO_3 \rightleftharpoons KAg(CN)_2 + KNO_3$

6. 1 litre of 0·1 M NaCl is added to 5 litres of 0·1 M NaBr and 4 litres of 0·1 M $BaCl_2$ so that the total volume becomes 10 litres.
   Which statements follow?
   a. $[Br^-]$ is 5×10^{-2} mol l^{-1}
   b. $[Na^+]$ is 6×10^{-2} mol l^{-1}
   c. $[BaCl_2]$ is 4×10^{-2} mol l^{-1}
   d. $[Ba^{2+}]$ is 4×10^{-2} mol l^{-1}
   e. $[Cl^-]$ is 9×10^{-2} mol l^{-1}

IONIC EQUILIBRIUM

7. One drop of M $AgNO_3$ is added to water and the solution made up to 1 litre. If we assume that one drop is 0·05 ml and that the solubility product of AgCl is 10^{-10}, then

   a. the drop is diluted 20 000 times.
   b. $[AgNO_3]$ is 2×10^{-4} mol l^{-1}.
   c. (Ag^+) is 2×10^{-5} mol l^{-1}.
   d. (NO_3^-) is 5×10^{-5} mol l^{-1}.
   e. the solution would give a precipitate with M HCl.

8. Which of the following would readily give a precipitate of silver chloride when added to a litre of 0·01 M $AgNO_3$?

   a. One drop of conc. HCl.
   b. One drop of acetyl chloride.
   c. One drop of phosphorus trichloride.
   d. One drop of thionyl chloride.
   e. One drop of carbon tetrachloride.

9. Which of the following solutions would readily dissolve a precipitate of silver chloride with the formation of a complex ion?

   a. 880 ammonia.
   b. M NaOH.
   c. M KCN.
   d. M $Na_2S_2O_3$.
   e. M KI.

10. The solubility of $RaSO_4$ is given as $6·5 \times 10^{-8}$ mol l^{-1} at 25°C. The solubility product of $BaSO_4$ is 1×10^{-10}. Which statements follow?

    a. Solubility product of $RaSO_4$ is $4·2 \times 10^{-17}$.
    b. Solubility product of $RaSO_4$ is greater than that of $BaSO_4$.
    c. $RaSO_4$ is more insoluble than $BaSO_4$ (values in mol l^{-1}).
    d. Solubility products of both salts would probably be greater at 50°C.
    e. Given saturated solutions of both salts, $[SO_4^{2-}]$ is greater in the $RaSO_4$ solution.

11. The solubility of CaF_2 is given as 0·0002 mol l^{-1} at 25°C. Which statements follow?

    a. Solubility is 2×10^{-4} mol l^{-1}.
    b. Solubility is 0·0118 g l^{-1}.
    c. In a saturated solution, $[Ca^{2+}]$ is 2×10^{-4} mol l^{-1}.
    d. In a saturated solution, $[F^-]$ is 2×10^{-4} mol l^{-1}.
    e. Solubility product is 8×10^{-12}

IONIC EQUILIBRIUM

12. Concentrated sulphuric acid is added to a suspension of $BaSO_4$ in water. (Neglect any rise in temperature.) Which statements follow?

   a. $[SO_4^{2-}]$ is increased.
   b. The solubility product of $BaSO_4$ is increased.
   c. The ionic product $[Ba^{2+}][SO_4^{2-}]$ becomes greater than the solubility product.
   d. The suspension begins to dissolve.
   e. The precipitate increases.

13. The following dissociation constants (or pK values) were taken from a book of reference. Which is the strongest acid?

   a. HCN $pK = 9.40$
   b. HF $K = 8.7 \times 10^{-4}$
   c. HNO_2 $K = 8.1 \times 10^{-4}$
   d. Formic acid $pK = 3.75$
   e. Acetic acid $pK = 4.76$

14. The following solubility products were taken from a book of reference. Which of the compounds has the lowest solubility in mol l^{-1}.

   a. AgCl 1.56×10^{-10}
   b. AgBr 7.7×10^{-13}
   c. AgCNS 1.2×10^{-12}
   d. Ag_2CrO_4 2.4×10^{-12}
   e. Ag_3PO_4 1.8×10^{-18}

15. The following stability constants (or log K values) were taken from a book of reference. Which appears to be the most stable complex ion?

   a. $[CdI_4]^{2-}$ $K = 2 \times 10^6$
   b. $[Cd(NH_3)_4]^{2+}$ $\log K = 6.6$
   c. $[Co(NH_3)_6]^{2+}$ $\log K = 4.9$
   d. $[Ag(NH_3)_2]^+$ $K = 1.7 \times 10^7$
   e. $[Ni(NH_3)_6]^{2+}$ $K = 4.8 \times 10^7$

16. Which of the following changes involve a change in coordination number as well as a change of ligand?

   a. $[Co(H_2O)_6]^{2+} \rightleftharpoons [CoCl_4]^{2-}$
   b. $[Cu(H_2O)_4]^{2+} \rightleftharpoons [Cu(NH_3)_4]^{2+}$
   c. $[Co(H_2O)_6]^{2+} \rightleftharpoons [Co(NH_3)_6]^{3+}$
   d. $[Fe(H_2O)_6]^{2+} \rightleftharpoons [Fe(CN)_6]^{4-}$
   e. $[Cd(H_2O)_4]^{2+} \rightleftharpoons [CdI_4]^{2-}$

17. Which of the following changes involve a change in oxidation state as well as a change of ligand?

   a. $[Ni(H_2O)_6]^{2+} \rightleftharpoons [Ni(NH_3)_6]^{2+}$
   b. $Fe^{3+}(aq) \rightleftharpoons Fe(CNS)^{2+}(aq)$
   c. $[Fe(H_2O)_6]^{2+} \rightleftharpoons [Fe(CN)_6]^{4-}$
   d. $[Cu(H_2O)_4]^{2+} \rightleftharpoons [Cu(CN)_4]^{3-}$
   e. $[Co(H_2O)_6]^{2+} \rightleftharpoons [CoCl_4]^{2-}$

IONIC EQUILIBRIUM

18. Dilute hydrochloric acid is added to a solution of H_2S so that the pH becomes 3. This solution is now added in excess to molar solutions of Cd^{2+} and Mn^{2+} in separate test tubes. Hence
(Assume that for a saturated solution of H_2S
$[H^+]^2[S^{2-}] = 10^{-23}$
Solubility product of CdS 5×10^{-25}
Solubility product of MnS 1×10^{-15})

   a. $[S^{2-}]$ in this acidified solution is 10^{-29}.
   b. CdS is precipitated.
   c. MnS is precipitated.
   d. $[Mn^{2+}]$ left in solution is less than 10^{-3} mol l^{-1}.
   e. $[Cd^{2+}]$ left in solution is less than 10^{-6} mol l^{-1}.

19. Solid ammonium chloride is added to a solution of ammonium hydroxide so that $[OH^-]$ is about 10^{-6} mol l^{-1}.
Which of the following hydroxides would you expect to be precipitated when this ammonia solution is added to M/10 solutions of the metal ions?

	Solubility product
a. $Ca(OH)_2$	8×10^{-6}
b. $Mg(OH)_2$	3×10^{-11}
c. $Zn(OH)_2$	1×10^{-17}
d. $Cr(OH)_2$	3×10^{-29}
e. $Fe(OH)_2$	4×10^{-38}

20. In spite of the great value of the solubility product in predicting what takes place in precipitation reactions, simple calculations often suggest results which are not achieved in practice. Which of the following do you think are contributory factors?

   a. There is much uncertainty in many of the values quoted in reference books.
   b. Solutions become supersaturated if no solid is present.
   c. Values are quoted at 25°C and are often very different at other temperatures.
   d. Precipitates can absorb other ions which would not otherwise be precipitated.
   e. 'Activities' of ions should be used instead of 'concentration in mol l^{-1}'.

ELECTROCHEMISTRY

You may need the following approximate values:
1 faraday = 9.6×10^4 coulombs or 1 mol of electrons
Charge on an electron = 1.6×10^{-19} coulombs
Avogadro constant = 6.0×10^{23} mol^{-1}

Some standard electrode potentials: $E°$ (volts)

$K^+ + e^- \rightleftharpoons K$ − 2·93	$Cu^{2+} + 2e^- \rightleftharpoons Cu$	+ 0·34
$Na^+ + e^- \rightleftharpoons Na$ − 2·71	$O_2 + 2H_2O + 4e^- \rightleftharpoons 4OH^-$	+ 0·40
$Mg^{2+} + 2e^- \rightleftharpoons Mg$ − 2·37	$I_2 + 2e^- \rightleftharpoons 2I^-$	+ 0·54
$Al^{3+} + 3e^- \rightleftharpoons Al$ − 1·66	$Fe^{3+} + e^- \rightleftharpoons Fe^{2+}$	+ 0·77
$Zn^{2+} + 2e^- \rightleftharpoons Zn$ − 0·76	$Ag^+ + e^- \rightleftharpoons Ag$	+ 0·80
$Fe^{2+} + 2e^- \rightleftharpoons Fe$ − 0·44	$Br_2 + 2e^- \rightleftharpoons 2Br^-$	+ 1·07
$Ni^{2+} + 2e^- \rightleftharpoons Ni$ − 0·24	$O_2 + 4H^+ + 4e^- \rightleftharpoons 2H_2O$	+ 1·23
$2H^+ + 2e^- \rightleftharpoons H_2$ 0·00	$Cr_2O_7^{2-} + 14H^+ + 6e^- \rightleftharpoons 2Cr^{3+} + 7H_2O$	+ 1·33
$Sn^{4+} + 2e^- \rightleftharpoons Sn^{2+}$ + 0·14	$2H^+ + H_2O_2 + 2e^- \rightleftharpoons 2H_2O$	+ 1·77

Unless otherwise stated it is assumed that there is no appreciable change in concentration during the course of any of the reactions.

1. When 1 faraday of electricity is passed through a dilute solution of sulphuric acid it liberates
   a. 1 g-atom of hydrogen.
   b. 6×10^{23} atoms of hydrogen.
   c. ½ mole of oxygen gas.
   d. 8 g of oxygen.
   e. 3×10^{23} molecules of oxygen.

2. If 2 amperes of current are passed through a dilute solution of sulphuric acid with platinum electrodes for 4 hours the quantity of electricity which passes is
   a. 8 ampere-hours.
   b. 480 coulombs.
   c. 3 faradays.
   d. 1.8×10^{24} electrons.
   e. 8 watts.

3. 5·6 dm³ of hydrogen gas (at s.t.p.) were liberated during the electrolysis of dilute sodium hydroxide with platinum electrodes after a steady current of electricity had passed for forty minutes. Which statements are correct?
   a. 0·25 mole of hydrogen gas was evolved.
   b. 0·25 faraday was passed.
   c. 2.4×10^4 coulombs were passed.
   d. The current was 10 amperes.
   e. The total volume of electrolytic gas was 11·2 dm³.

4. 1 ampere is passed for 2 hours 40 minutes through separate solutions of copper

51

sulphate, lead nitrate, and silver nitrate. Which statements are correct?

a. 1 faraday is used.
b. 1 mol of metal is plated from each of the solutions.
c. 20·7 g of lead are plated.
d. 63·5 g of copper are plated.
e. 10·8 g of silver are plated.

5. 5 volts are applied to platinum electrodes in a dilute solution of sulphuric acid. The result is that

a. hydrogen is liberated at the cathode.
b. oxygen is liberated at the anode.
c. the solution becomes more dilute.
d. the hydrogen ion concentration increases.
e. the pH increases.

6. An anode is always

a. the positive electrode.
b. the electrode at which current leaves a cell.
c. the electrode at which electrons leave a cell.
d. the electrode at which oxidation takes place.
e. the negative electrode.

7. Which of the following aqueous solutions would you expect to oxidise potassium iodide to iodine but not potassium bromide to bromine?

a. M ferric chloride
b. M ferrous sulphate
c. M stannous chloride
d. M potassium dichromate (acid)
e. M hydrogen peroxide (acid)

8. The ion which is discharged at the cathode during the electrolysis of an aqueous solution of two electrolytes (e.g. $CaCl_2$ and NaCl) depends on the

a. electrode (redox) potentials of the cations.
b. concentration of the cations.
c. temperature.
d. nature of the cathode.
e. current density.

9. If an aqueous solution of sodium sulphate is electrolysed (with platinum electrodes) one would expect to find

a. sodium deposited at the cathode.
b. oxygen liberated at the anode.
c. the solution around the anode becoming acidic.
d. the solution around the cathode becoming alkaline.
e. the solution as a whole becoming more dilute.

ELECTROCHEMISTRY

10. If an aqueous solution of acidified potassium iodide is electrolysed (with platinum electrodes) one would expect to find
    a. potassium deposited at the cathode.
    b. oxygen liberated at the anode.
    c. the solution around the cathode becoming alkaline.
    d. the solution around the anode becoming brown.
    e. iodine released at the cathode.

11. The e.m.f. of a Daniell cell
    $Zn \mid Zn^{2+} \parallel Cu^{2+} \mid Cu$
    can be increased by
    a. increasing the area of the electrodes.
    b. increasing the concentration of the copper sulphate solution.
    c. increasing the concentration of the zinc sulphate solution.
    d. using dilute sulphuric acid instead of copper sulphate.
    e. using dilute sulphuric acid instead of zinc sulphate.

12. Which of the following cells would you expect to produce an e.m.f. of more than 1·5 volts? (Assume molar aqueous solutions.)
    a. $Mg \mid Mg^{2+} \parallel Zn^{2+} \mid Zn$
    b. $Mg \mid Mg^{2+} \parallel Fe^{2+} \mid Fe$
    c. $Mg \mid Mg^{2+} \parallel Ni^{2+} \mid Ni$
    d. $Fe \mid Fe^{2+} \parallel Ag^+ \mid Ag$
    e. $Al \mid Al^{3+} \parallel Ni^{2+} \mid Ni$

13. A piece of pure zinc is placed in dilute sulphuric acid and the reaction is very slow. When a piece of silver is put in the acid and allowed to touch the zinc then
    a. the zinc dissolves rapidly.
    b. the zinc forms the cathode of a cell.
    c. the zinc metal is reduced.
    d. the silver metal is reduced.
    e. bubbles of hydrogen come from the silver.

14. Given that $E = E^\circ - 0\cdot06(pH)$ for a hydrogen electrode. If E is found to be $-0\cdot6$ volt when the electrode is placed in a given solution then the
    a. pH is zero.
    b. pH is 10.
    c. solution is 0·1 N acid.
    d. solution is alkaline.
    e. solution could be 10^{-4} M NaOH.

15. Given that $E = E^\circ + 0\cdot06 \log[Ag^+]$ for a silver electrode in solutions of silver ions then
    a. for 0·1 M $AgNO_3$, $E = 0\cdot74$ volt.
    b. for 10^{-4} M $AgNO_3$, $E = 0\cdot56$ volt.
    c. if $E = 0\cdot50$ volt for a saturated solution of AgCl, the solubility of AgCl is 10^{-5} mol l^{-1}.

d. the solubility product of AgCl is 10^{-10}.
e. if the solubility product of AgBr is 5×10^{-13}, the value of E for a saturated solution of AgBr is 0·35 volt.

16. Oxygen is not always liberated at the anode when sulphuric acid is electrolysed. The product is dependent upon the

a. nature of the anode.
b. nature of the cathode.
c. temperature of the electrolyte.
d. current density.
e. concentration of the acid.

17. From an examination of the relevant standard potentials one might expect hydrogen to be liberated at the cathode of the Castner-Kellner flowing mercury cell, but it is not. Which of the following seem to be contributory causes?

a. A high concentration of H^+ ions causes E for $2H^+ + 2e^- \rightleftharpoons H_2$ to become negative.
b. A high concentration of OH^- ions causes E for $2H^+ + 2e^- \rightleftharpoons H_2$ to become negative.
c. There is a large hydrogen overvoltage at a mercury electrode.
d. Sodium is not liberated as the metal but as sodium amalgam. This requires a lower negative potential.
e. Sodium amalgam does not react with aqueous sodium chloride to give hydrogen.

18. Confusion sometimes exists between the ionisation potential (I°) and the electrode potential (E°) for the change
$Na \rightarrow Na^+ + e^-$
Which statements are correct?

a. I° is the energy needed to remove an electron from a gaseous atom of sodium.
b. E° refers to hydrated ions in aqueous solution.
c. E° refers to ions of sodium in the fused state.
d. E° is better defined as the potential for the change
$Na^+(aq) + e^- \rightarrow Na(s)$
e. I° is better defined as the potential required to bring about the change
$Na(g) \rightarrow Na^+(g) + e^-$

19. Which of the following instruments make use of the standard electrode potentials and the variation of electrode potential with concentration?

a. Polarimeter
b. Polarograph
c. pH meter
d. Electrophorus
e. Electrophotometer

20. i. The stability constant for the reaction
$$Ag^+ + 2(CN)^- \rightleftharpoons Ag(CN)_2^-$$
is given as 10^{21}.

ii. $E = E^\circ + 0\cdot06 \log[Ag^+]$

iii. $E^\circ = +0\cdot8$ volt

An equal volume of 2 M KCN is added to 0·2 M $AgNO_3$ to provide an electroplating solution. If reasonable approximations are made then

a. there is a large excess of cyanide ions in the mixture.

b. $[CN^-]$ is about 1 mol l^{-1}.

c. $[Ag(CN)_2^-]$ is about 10^{-1} mol l^{-1}.

d. $[Ag^+]$ is about 10^{-20} mol l^{-1}.

e. the electrode potential (E) is now about $+0\cdot5$ volt.

ENERGETICS

1. $H_2S(g) + 1\frac{1}{2}O_2(g) \rightarrow H_2O(l) + SO_2(g)$
   $\Delta H = -565$ kJ
   This statement implies that

   a. the reaction is endothermic.
   b. 565 kJ are taken in when 1 mole of H_2S gas is burned in excess oxygen.
   c. the reaction is carried out at constant pressure.
   d. the enthalpy of the system decreases.
   e. volumes of gases are corrected to s.t.p.

2. The value of ΔH given for the reaction in Question 1 is more precisely stated in advanced books as ΔH°. This implies:

   a. Pressure of 1 atmosphere
   b. Constant volume
   c. Reactants and products are in the states quoted (standard states)
   d. 25°C
   e. 273K

3. Heats of combustion $(-\Delta H)$ are determined in a bomb calorimeter. Which statements are correct?

   a. The heat given out is actually measured at constant volume.
   b. The energy change at constant volume is ΔU.
   c. The difference between ΔH and ΔU is due to the work done by the expansion of gases.
   d. ΔH is more useful than ΔU since most changes take place at constant pressure.
   e. Values of ΔU are not usually quoted in elementary textbooks.

4. The heat of formation ΔH_f of a compound is always quoted for the normal form of the element at 25°C and at atmospheric pressure. Which of the following would give acceptable values for ΔH_f?

   a. $S(rhombic) + O_2(g) \rightarrow SO_2(g)$
   b. $S(monoclinic) + O_2(g) \rightarrow SO_2(g)$
   c. $C(graphite) + O_2(g) \rightarrow CO_2(g)$
   d. $C(diamond) + O_2(g) \rightarrow CO_2(g)$
   e. $P_4(white) + 5O_2(g) \rightarrow P_4O_{10}(g)$

5. $C(s) + O_2(g) \rightarrow CO_2(g): \Delta H = -394$ kJ
   $S(s) + O_2(g) \rightarrow SO_2(g): \Delta H = -297$ kJ
   $CS_2(l) + 3O_2(g) \rightarrow CO_2(g) + 2SO_2(g):$
   $\Delta H = -1076$ kJ
   Which statements are correct?

   a. System $C(s) + O_2(g)$ loses 394 kJ.
   b. System $S(s) + O_2(g)$ loses 297 kJ.
   c. Total loss in making 1 mole of CO_2 and 2 moles of SO_2 from the elements is 988 kJ.
   d. ΔH_f of $CS_2(l)$ is -88 kJ.
   e. It would be impossible to make CS_2 by heating carbon and sulphur.

ENERGETICS

6. Methanol can be synthesised by the reaction:
$$2H_2(g) + CO(g) \rightarrow CH_3OH(l)$$
The heat of combustion of methanol is 715 kJ.
ΔH_f for $CO(g)$ is -109 kJ.
ΔH_f for $CO_2(g)$ is -394 kJ.
ΔH_f for $H_2O(l)$ is -286 kJ.
Which statements follow?

a. Heat given out when 2 moles of hydrogen gas are converted to 2 moles of water is 572 kJ.
b. $CO(g) + \frac{1}{2}O_2(g) \rightarrow CO_2(g)$:
$\Delta H = -285$ kJ.
c. Total heat of combustion of $2H_2(g)$ and $CO(g)$ is 857 kJ.
d. ΔH for the synthesis of methanol is 142 kJ mol^{-1}.
e. Applying Le Chatelier's Principle, a high temperature would tend to produce an increased proportion of methanol in the equilibrium.

7. Acetylene can be produced according to the equation:
$$2CH_4(g) \rightarrow C_2H_2(g) + 3H_2(g).$$
Given:
i. $CH_4(g) + 2O_2(g) \rightarrow CO_2(g) + 2H_2O(l)$:
$\Delta H = -882$ kJ.
ii. $C_2H_2(g) + 2\frac{1}{2}O_2(g) \rightarrow 2CO_2(g) + H_2O(l)$
$(l) \Delta H = -1306$ kJ.
iii. ΔH_f for $H_2O(l)$ is -286 kJ
Then:

a. combustion of 2 moles of methane gives out 1764 kJ.
b. combustion of 3 moles of hydrogen produces 858 kJ.
c. more heat is given out from the combustion of 2 moles of methane than from 1 mole of acetylene and 3 moles of hydrogen.
d. ΔH for the formation of acetylene from methane is -400 kJ.
e. the production of acetylene from methane is favoured by a high temperature.

8. Formaldehyde can be produced by the oxidation of methanol
i. $CH_3OH(l) + \frac{1}{2}O_2(g) \rightarrow CH_2O(g) + H_2O(l)$
or by the dehydrogenation of methanol
ii. $CH_3OH \rightarrow CH_2O + H_2$
If ΔH_f for $CH_3OH(l) = -239$ kJ.
ΔH_f for $CH_2O(g) = -116$ kJ.
ΔH_f for $H_2O(l) = -286$ kJ.
Then:

a. gain in enthalpy in system (i) is 163 kJ.
b. gain in enthalpy in system (ii) is 123 kJ.
c. since (i) is exothermic it should be the more economical process.
d. it might be possible to combine the two methods by balancing the exothermic and endothermic reactions.
e. the suggestion in (d) is actually used.

9. Given the information that
$$C_6H_6(l) + 7\frac{1}{2}O_2(g) \rightarrow 6CO_2(g) + 3H_2O(l):$$
$$\Delta H = -3273 \text{ kJ}$$
$$C_2H_2(g) + 2\frac{1}{2}O_2(g) \rightarrow 2CO_2(g) + H_2O(l):$$
$$\Delta H = -1306 \text{ kJ}$$

Then:

a. when 3 moles of acetylene are oxidized 3918 kJ are evolved.
b. $3C_2H_2(g) \rightarrow C_6H_6(l): \Delta H = -645 kJ$.
c. the synthesis of benzene from acetylene is an exothermic process.
d. application of Le Chatelier's Principle suggests a low temperature for the synthesis of benzene.
e. it follows that acetylene can be converted to benzene without initial heating.

10. Given:
$C(graphite) + 2H_2(g) \rightarrow CH_4(g):$
$\Delta H = -73$ kJ.
$C(graphite) \rightarrow C(free\ atom):$
$\Delta H = +715$ kJ.
$\frac{1}{2}H_2(g) \rightarrow H(atom): \Delta H = +218$ kJ
Then:

a. $2H_2(g) \rightarrow 4H(atom): \Delta H = +872$ kJ.
b. $C(graphite) + 2H_2 \rightarrow C(atom) + 4H(atom):$
$\Delta H = +1587$ kJ
c. $C(atom) + 4H(atom) \rightarrow CH_4(gas):$
$\Delta H = -1660$ kJ
d. since there are four equivalent bonds in methane, the average heat of formation of each is -415 kJ.
e. 415 kJ would be needed to break the C—H bond.

11. Given:
i. $2C(graphite) + 2H_2(g) \rightarrow C_2H_4(g):$
$\Delta H = +52$ kJ
ii. $\frac{1}{2}H_2(g) \rightarrow H(atom): \Delta H = +218$ kJ
iii. $C(graphite) \rightarrow C(atom):$
$\Delta H = +715$ kJ
iv. Total bond energy of methane = 1660 kJ
Which of the following steps in calculating the C=C bond energy are true?

a. 1430 kJ are needed to convert 2 moles of graphite to atoms.
b. 872 kJ are needed to convert 2 moles of hydrogen gas to atoms.
c. 2302 kJ are needed to convert 2 moles of graphite and 2 moles of hydrogen gas to atoms.
d. The total bond energy of the C=C and 4 C—H bonds is 2354 kJ.
e. The bond energy of the C=C bond is 590 kJ.

For the next three questions (12, 13, and 14) the following information is required to investigate the nature of the bonds in benzene.

ENERGETICS

i. $\begin{cases} C_6H_6(l) + 7\tfrac{1}{2}O_2(g) \rightarrow 6CO_2(g) \\ \qquad\qquad\qquad\qquad + 3H_2O(l) \\ \text{Combustion of benzene} \\ \qquad\qquad \Delta H = -3273 \text{ kJ} \end{cases}$

ii. $\begin{cases} C(s) + O_2(g) \rightarrow CO_2(g) \\ \text{Formation of } CO_2 \; \Delta H_f = -394 \text{ kJ} \end{cases}$

iii. $\begin{cases} H_2(g) + \tfrac{1}{2}O_2(g) \rightarrow H_2O(l) \\ \text{Formation of } H_2O(l) \; \Delta H_f = -286 \text{ kJ} \end{cases}$

iv. $\begin{cases} H_2(g) \rightarrow 2H(g) \\ \text{Atomisation of } H_2(g) \; \Delta H = +436 \text{ kJ} \end{cases}$

v. $\begin{cases} C(s) \rightarrow C(g) \\ \text{Atomisation of graphite} \\ \qquad\qquad \Delta H = +715 \text{ kJ} \end{cases}$

Bond energies —C—H = 415 kJ
—C—C— = 348 kJ
—C=C— = 590 kJ

Which statements are correct?

12.
a. $6C(s) + 6O_2(g) \rightarrow 6CO_2: \Delta H = -2364$ kJ
b. $3H_2(g) + 1\tfrac{1}{2}O_2 \rightarrow 3H_2O: \Delta H = -858$ kJ
c. Total heat of combustion of $6C(s)$ and $3H_2(g): \Delta H = -3222$ kJ
d. Heat of formation of benzene:
$\Delta H_f = -51$ kJ
e. Benzene is readily synthesized by heating carbon in hydrogen.

13.
a. $6C(s) \rightarrow 6C(g) \quad \Delta H = +4290$ kJ
b. $3H_2(g) \rightarrow 6H(g) \; \Delta H = +1308$ kJ
c. Total heat of atomisation of $6C(s)$ and $3H_2(g): \Delta H = +5598$ kJ
d. Sum of all the bond energies in benzene is $+5598$ kJ
e. The sum of the energies of 6 carbon–hydrogen and 6 carbon–carbon bonds is 5547 kJ.

14.
a. Sum of 6 C—H bond energies is 2490 kJ.
b. Sum of the remaining bonds is 3057 kJ.
c. Average bond energy of the carbon-to-carbon link is 509 kJ.
d. This is between the energies of the C—C bond and the C=C bond.
e. The bond energy agrees with 3 C—C bonds and 3 C=C bonds.

15. The following definitions and quantities appeared in a book of chemical constants:
Lattice energy, the energy to change 1 mole of compound from the crystalline state to gaseous ions.
Hydration energy, the energy to change 1 mole of compound from aqueous solution to gaseous ions.

	Lattice energy	Hydration energy
LiBr	800	854 kJ
KBr	670	557 kJ

Which statements are correct?

a. $LiBr(s) \rightarrow Li^+(aq) + Br^-(aq):$
$\Delta H = -54$ kJ.
b. $KBr(s) \rightarrow K^+(aq) + Br^-(aq):$
$\Delta H = -113$ kJ.
c. One would expect a rise in temperature when solid LiBr was dissolved in water.
d. One would expect a fall in temperature when solid KBr was dissolved in water.
e. One might expect the difference to be due to the water molecules being held closer to the smaller lithium ions.

16. *The Electron affinity* of an element is quoted as the energy liberated when an electron is received by an atom.
e.g. $Cl(g) \rightarrow Cl^-(g)$: $\Delta H = -365$ kJ
We are given that:

i. {Energy of sublimation
   $Na(s) \rightarrow Na(g)$: $\Delta H = +109$ kJ

ii. {$\frac{1}{2}$ Energy of dissociation
   $\frac{1}{2}Br_2(l) \rightarrow Br(g)$: $\Delta H = +97$ kJ

iii. {Energy of ionisation
    $Na(g) \rightarrow Na^+(g)$: $\Delta H = +496$ kJ

iv. {Lattice energy
    $NaBr(s) \rightarrow Na^+(g) + Br^-(g)$:
    $\Delta H = +718$ kJ

v. {Heat of formation
   $Na(s) + \frac{1}{2}Br_2(l) \rightarrow NaBr(s)$:
   $\Delta H = -361$ kJ

Which statements follow?

a. $Na(s) + \frac{1}{2}Br_2(l) \rightarrow Na^+(g) + Br^-(g)$:
$\Delta H = +357$ kJ
b. $Na(s) + \frac{1}{2}Br_2(l) \rightarrow Na(g) + Br(g)$:
$\Delta H = +206$ kJ
c. $Na(s) + \frac{1}{2}Br_2(l) \rightarrow Na^+(g) + Br(g)$:
$\Delta H = +702$ kJ
d. $Br(g) \rightarrow Br^-(g)$: $\Delta H = -345$ kJ
e. Energy is liberated when a bromide ion is formed from a bromine atom.

17. 7·0 g of copper powder were added to 200 ml of M $AgNO_3$ with continual stirring, in a plastic bottle. The temperature rose rapidly from 15·0°C to 32·6°C. If we assume
   i. no heat losses,
   ii. the container has a negligible heat capacity,
   iii. the heat capacity of the solution is the same as that of water (4·2 J g^{-1} K^{-1}),
which of the following statements are correct?

a. If 14·0 g of copper had been used the temperature would have risen to 50·2°C.
b. If 2 litres of 0·1 M $AgNO_3$ had been used the temperature rise would have been 17·6K.
c. The energy liberated is 148 kJ mol^{-1} of silver displaced.
d. If we had started with 200 ml of water at 15°C, dissolved sufficient $AgNO_3$ to make a molar solution and then immediately added 7·0 g of copper, the same temperature rise would have been recorded.
e. $Cu(s) + 2Ag^+(aq) \rightarrow 2Ag(s) + Cu^{2+}(aq)$
$\Delta H = -148$ kJ

18. A primary cell is constructed from a silver electrode in M $AgNO_3$ and a copper electrode in M $CuSO_4$, joined by the usual salt bridge. Potentiometer readings give an e.m.f. of 0·45 volt. If a very small current is taken from the cell so that 0·1

mole of Cu is dissolved, which of the following statements are true?

a. 0·1 faraday is transferred.
b. 86 850 joules of electrical energy are produced.
c. The maximum available work from the reaction is 86·85 kJ mol^{-1} of copper used up.
d. $Cu(s) + 2Ag^+(aq) \rightarrow 2Ag(s) + Cu^{2+}(aq)$
$\Delta G = -86·85$ kJ
e. The e.m.f. will be doubled if 2 M solutions are used.

19. In the reaction
$Cu(s) + 2Ag^+(aq) \rightarrow Cu^{2+}(aq) + 2Ag(s)$
if we accept the values of ΔH and ΔG we have calculated in Questions 17 and 18, then:

a. the enthalpy of the system decreases.
b. the free energy of the reaction decreases by an even larger amount.
c. the heat given out in this reaction is greater than the maximum useful electrical energy.
d. the loss in free energy indicates that the reaction will proceed spontaneously.
e. the difference in ΔH and ΔG is due to the change in entropy (or disorder) of the system.

20. If the values of ΔG_f (the free energy of formation) taken from a book of physical and chemical constants are
$CO = -137$ kJ mol^{-1}
$CO_2 = -395$ kJ mol^{-1}
$H_2O = -237$ kJ mol^{-1}
$CH_3OH = -166$ kJ mol^{-1}
$HCHO = -110$ kJ mol^{-1}
$HCO_2H = -346$ kJ mol^{-1}
which of the following reactions would appear to be possible?

a. $CO + 2H_2 \rightarrow CH_3OH$
b. $CO_2 + H_2 \rightarrow HCO_2H$
c. $CO + H_2O \rightarrow HCO_2H$
d. $CO + H_2 \rightarrow HCHO$
e. $C + H_2O \rightarrow HCHO$

PHASE EQUILIBRIA

1. A melt, consisting of a mixture of two metals, freezes suddenly when cooled to 420°C. This is consistent with the formation of
   a. a eutectic mixture of the metals.
   b. a eutectic mixture of one of the metals and a compound which is formed.
   c. a compound.
   d. a solid solution.
   e. a cryohydrate.

2. When a melt of two metals is cooled slowly
   a. there is a sudden change in volume when the solid starts to form.
   b. the rate of cooling decreases when the solid starts to form.
   c. one metal at a time starts to crystallise out, no matter what the metals are.
   d. it is quite possible that the temperature may rise as solid forms.
   e. the solid which forms must be of the same composition as the liquid when a solid solution starts to crystallise from a liquid.

3. Transition elements readily form solid solutions. Which of the following do you think are valid reasons for this?
   a. Atoms of the elements have similar radii.
   b. Atoms of the elements have similar atomic weights.
   c. Atoms of the elements have similar charge to mass ratios.
   d. The elements are in the same period of the periodic table.
   e. The elements are in the same group of the periodic table.

4. Which of the following metal pairs are likely to form solid solutions?
   a. Zinc and copper.
   b. Copper and nickel.
   c. Platinum and gold.
   d. Tin and lead.
   e. Cadmium and bismuth.

In answering Questions 5–12 you are required to study the diagrams overleaf and choose reasonable answers. It may help if you make your own copies of the diagrams on graph paper.

PHASE EQUILIBRIA

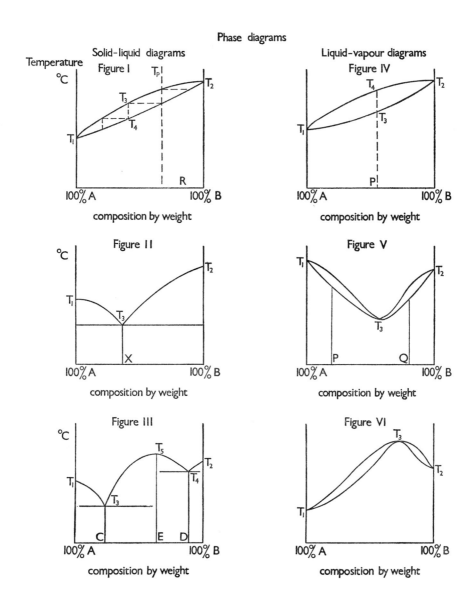

Phase diagrams

PHASE EQUILIBRIA

5. Use figure I.
 A is copper m.p.(T_1) = 1090°C
 B is nickel m.p.(T_2) = 1450°C
 A melt of 30 per cent copper and 70 per cent nickel is cooled from temperature T_P.

 a. The solid which first separates is copper.
 b. The solid which first separates contains about 80 per cent copper.
 c. The solid which finally separates is richer in copper than that which first separates.
 d. The melt finally solidifies at temperature T_3.
 e. The melt finally solidifies at temperature T_4.

6. Use figure II.
 A mixture of tin and lead is cooled.
 A is tin m.p.(T_1) = 232°C
 B is lead m.p.(T_2) = 326°C
  T_3 = 183°C
 The composition of X is 63 per cent tin and 37 per cent lead.

 a. If a melt of equal weights of tin and lead is cooled it will freeze completely at 200°C.
 b. If a melt of equal weights of tin and lead is cooled it will become pasty between 200°C and 183°C.
 c. If a melt of 2/3 lead and 1/3 tin is cooled, it will be pasty from 240°C to 183°C.
 d. If a melt of 1/3 lead and 2/3 tin is cooled, it will freeze suddenly at 183°C.
 e. Composition X makes the best solder for electrical components.

7. Use figure III.
 A is tin m.p.(T_1) = 232°C
 B is magnesium m.p.(T_2) = 640°C
 C is 10 per cent magnesium and 90 per cent tin,
 Temperature T_3 = 210°C.
 D is 90 per cent magnesium and 10 per cent tin,
 Temperature T_4 is 560°C,
 Temperature T_5 is 800°C,
 E is 30 per cent magnesium and 70 per cent tin.

 a. The maximum value indicates that composition E is in accordance with a simple compound of tin and magnesium.
 b. The formula of the compound between tin and magnesium is $MgSn_2$.
 c. T_3 and T_4 are the temperatures at which eutectic mixtures separate.
 d. If a molten mixture of 50 per cent of each component is cooled, it starts to solidify at 700°C and is completely solid at 560°C.
 e. Magnesium solidifies first from a 50 per cent molten mixture.

8. Use figure III.
 A is phenol m.p.(T_1) = 40°C
 B is 4-amino toluene m.p.(T_2) = 43°C
 C is 76 per cent phenol and the minimum temperature T_3 is 8°C.
 D is 31 per cent phenol and the minimum temperature T_4 is 20°C.

The maximum temperature T_5 is 28°C.
E is 47 per cent phenol and 53 per cent 4-amino toluene
If a 50 per cent liquid mixture is cooled from 50°C then:

a. solid 4-amino toluene first separates out.
b. a solid compound first separates of probable composition.

$$\left[CH_3-\bigcirc-NH_3\right]^+ \left[O-\bigcirc\right]^-$$

c. the mixture finally solidifies at 8°C.
d. the mixture which finally freezes is a eutectic mixture consisting of a compound of the two components, and 4-amino toluene.
e. there are only three mixtures of phenol and 4-amino toluene which have freezing points at the same temperatures as their melting points.

9. Use figure IV.
   A is hexane b.p. 68°C (T_1)
   B is octane b.p. 125°C (T_2)
   A mixture of these two components (composition P) is heated in a flask fitted with a Liebig condenser for distillation.

a. The mixture first boils at T_3.
b. The first distillate is hexane.
c. The first distillate is 80 per cent hexane.
d. The temperature of the liquid in the flask rises steadily but stops at T_4.
e. The final distillate is about 80 per cent hexane.

10. Use figure V.
    A is water b.p. 100°C (T_1)
    B is propanol b.p. 97°C (T_2)
    The azeotrope contains 28 per cent water and T_3 is 88°C.
    A 50 per cent mixture by weight is distilled with a still head.

a. The first distillate is pure water.
b. The first distillate is pure propanol.
c. The first distillate is the azeotropic mixture of 28 per cent water and 72 per cent propanol.
d. The final residue is nearly pure propanol.
e. The final residue is the azeotropic mixture of 28 per cent water and 72 per cent propanol.

11. Use figure V.
    A is toluene b.p. 110°C (T_1)
    B is ethanol b.p. 78°C (T_2)
    The azeotrope contains 68 per cent by weight of ethanol and T_3 is 76°C.
    A mixture containing 20 per cent ethanol is distilled with a still head.

a. The distillate is pure toluene.
b. The distillate is pure ethanol.
c. The distillate contains 68 per cent ethanol.
d. The final residue is pure toluene.
e. The final residue contains 32 per cent toluene.

12. Use figure VI.
    A is nitric acid b.p. 86°C (T_1)
    B is water b.p. 100°C (T_2)
    The azeotrope contains 68 per cent by weight of water.
    $T_3 = 120°C$
    A mixture containing 50 per cent water is distilled with a still head.

    a. The distillate contains 32 per cent nitric acid.
    b. The distillate which first distils over is pure nitric acid.
    c. The final residue in the flask is water.
    d. The final residue in the flask contains 32 per cent nitric acid.
    e. It is impossible to produce pure nitric acid from 50 per cent acid by such a process of distillation.

13. The freezing point curves for a nitric acid-water mixture show TWO maximum and THREE minimum values. The two maximum values occur at 77·8 per cent HNO_3 and 53·8 per cent HNO_3 respectively (by weight).
    Which statements follow?

    a. At the maximum and minimum values the composition of the liquid and of the solid in contact with it must be the same.
    b. The maximum values show the presence of compounds of nitric acid and water.
    c. One of the maximum values shows the presence of a compound $HNO_3, 3H_2O$.
    d. The three minimum values show the presence of three different eutectic mixtures.
    e. It should be possible to get pure nitric acid by freezing a dilute solution containing 60 per cent nitric acid.

14. If two immiscible liquids (e.g. water and aniline) are boiled together and the vapour condensed, then:

    a. the boiling point of the mixture is lower than either of the components.
    b. the total pressure of the boiling liquids is the same as the pressure of the atmosphere.
    c. the liquid which is formed from the condensed vapour contains a greater number of moles of the more volatile component than did the original mixture.
    d. the ratio of the weights of the two components is the same as the ratio of their vapour pressures at the temperature of the boiling mixture.
    e. the method of steam distillation is a good method of separating water from aniline.

15. The saturated vapour pressure of pure water at 60°C is 150 mmHg.
    Which of the following seem reasonable values?

    a. s.v.p. of water at 80°C is 70 mmHg
    b. s.v.p. of ether at 60°C is 500 mmHg
    c. s.v.p. of ethanol at 60°C is 500 mmHg
    d. s.v.p. of an aqueous solution of salt is 180 mmHg at 60°C
    e. s.v.p. of an aqueous solution of sugar is 160 mmHg at 60°C

PHASE EQUILIBRIA

16. Magnesium sulphate forms hydrates with 7, 6, 5, 4, 1 molecules of water of crystallisation, having dissociation pressures of 11·5, 10, 9, 4 and 1 mmHg respectively at 25°C.
If the vapour pressure of water in the atmosphere is

a. 5 mmHg, then Epsom salt effloresces.
b. 5 mmHg, then anhydrous magnesium sulphate gains water from the atmosphere.
c. 5 mmHg, then $MgSO_4, 5H_2O$ is the stable form.
d. 9·5 mmHg, then $MgSO_4, 6H_2O$ is the stable form.
e. 12 mmHg, then Epsom salt deliquesces.

17.

Salt hydrate	V.P. of hydrate	V.P. of saturated solution
$CuSO_4, 5H_2O$	5 mmHg	16 mmHg
$Na_2SO_4, 10H_2O$	16·3 mmHg	16·6 mmHg
$CaCl_2, 6H_2O$	2·5 mmHg	7·5 mmHg

The vapour pressure of the atmosphere on a certain date was measured in the given localities.
   3 mmHg in the Sahara Desert,
   20 mmHg in Malaya,
   10 mmHg in London.   (all at 25°C)
Which statements follow?

a. All the salts would deliquesce in Malaya.
b. Hydrated calcium chloride would deliquesce in all three localities.
c. Hydrated calcium chloride would become anhydrous in the Sahara Desert.
d. Copper sulphate crystals would deliquesce in London, but not in the Sahara.
e. A solution of copper sulphate saturated at 25°C would not readily form crystals in Malaya.

18. The Distribution (Partition) Law states that the ratio of concentration of solute in solvent A to the concentration of solute in solvent B is constant, provided that

a. the solvents are immiscible.
b. neither solvent is saturated with the solute.
c. the solute is in the same molecular form in both solvents.
d. the temperature is constant.
e. no other molecular species is present.

19. A weak organic acid is shaken with mixtures of carbon tetrachloride and water so that the following results are obtained.

Concentration of acid in water layer. (C_1)	Concentration of acid in CCl_4 layer. (C_2)
5 g/litre	0·2 g/litre
7 g/litre	0·4 g/litre
13 g/litre	1·4 g/litre

Which statements follow?

a. The value $\dfrac{C_1}{C_2}$ appears to be constant.
b. The value $\dfrac{C_1^2}{C_2}$ appears to be constant.
c. The value $\dfrac{\sqrt{C_1}}{C_2}$ appears to be constant.
d. The acid must be 100 per cent dissociated into ions in the water.
e. The acid is probably dimerised in the CCl_4 layer.

20. 5 ml 0·5 M $CuSO_4$(aq) were made up to 50 ml with ammonia solution and 50 ml of chloroform were added. If $\dfrac{\text{concentration of free } NH_3 \text{ in water}}{\text{concentration of free } NH_3 \text{ in } CHCl_3} = \dfrac{25}{1}$ and it needs 14 ml of 0·1 M HCl to neutralise the free NH_3 in the chloroform and 45 ml M HCl to neutralise the NH_3 in the water layer, then:

a. the cupric ions remain in the aqueous layer as complex ions with ammonia.
b. the total ammonia in the aqueous layer is $4·5 \times 10^{-2}$ mol.
c. the free ammonia in the aqueous layer is $3·5 \times 10^{-2}$ mol.
d. $1·0 \times 10^{-2}$ mol of ammonia is associated with $2·5 \times 10^{-3}$ mole of copper ions.
e. the copper ions are in the form of complex ions of formula $[Cu(NH_3)_2]^{2+}$.

STATES OF MATTER

1. The Kinetic Theory makes certain assumptions about gas molecules (or atoms in the case of monatomic gases). Which of the following are NOT necessary assumptions?

   a. The particles are in random motion with high velocities.
   b. Particles are small compared with the distances between them.
   c. Particles are perfectly elastic.
   d. All particles have the same kinetic energy at a given temperature.
   e. The kinetic energy of the particles increases with rising temperature.

2. From the Kinetic Theory of gases it is possible to derive the fundamental equation
   $$pv = 1/3\, L m \bar{u}^2 = RT$$
   Which of the following are true?

   a. $\bar{u}$ is the average velocity of the particles.
   b. $\tfrac{1}{2} m \bar{u}^2$ is the average kinetic energy of the particles.
   c. The total kinetic energy of the particles is $\tfrac{1}{2} N m \bar{u}^2$.
   d. The kinetic energy of the particles is directly proportional to the absolute temperature.
   e. L is the Avogadro constant.

3. If we accept the fundamental equation of the Kinetic Theory of gases
   $$pv = 1/3\, L m \bar{u}^2$$
   and we are given that the density (ρ) of hydrogen is 9×10^{-2} kg m^{-3} at atmospheric pressure of 10^5 N m^{-2} and temperature 273 K, then

   a. the density of hydrogen $\dfrac{Lm}{v}$ is $3p/\bar{u}^2$
   b. $\bar{u} = \sqrt{\dfrac{3p}{\rho}}$
   c. $\bar{u} = \sqrt{\left(\dfrac{3 \times 10^5 \text{ N m}^{-2}}{9 \times 10^{-2} \text{ kg m}^{-3}}\right)}$
   d. $\bar{u} = 1\cdot 8 \times 10^3$ m s^{-1}
   e. the r.m.s. velocity of the hydrogen molecules is more than 3000 m.p.h.

4. We are given 1 mole each of two gases A and B at the same temperature. If the r.m.s. values of the velocities are $\bar{u}_a$ and $\bar{u}_b$ and the masses are m_a and m_b respectively, then

   a. $m_a \bar{u}_a^2 = m_b \bar{u}_b^2$
   b. $\dfrac{m_a}{m_b} = \dfrac{\bar{u}_a^2}{\bar{u}_b^2}$
   c. $\dfrac{m_a}{m_b} = \dfrac{\bar{u}_a}{\bar{u}_b}$
   d. carbon monoxide diffuses through a porous plug faster than nitrogen.
   e. ^{235}UF$_6$ diffuses through a porous plug faster than ^{238}UF$_6$.

STATES OF MATTER

5. A toxic gas diffuses through a porous plug at about 1/7 of the speed of hydrogen gas at the same temperature and pressure. Then it follows that:

   a. the vapour density of the gas is about 7.
   b. the molar mass of the gas is about 50.
   c. the gas could be phosgene ($COCl_2$).
   d. oxygen diffuses at about $\frac{1}{4}$ of the speed of hydrogen.
   e. the toxic gas diffuses at about 4/7 of the speed of oxygen.

6. Differences between solids, liquids, and gases are explained in terms of the Kinetic Theory by which of the following statements?

   a. Particles are closer together in liquids than in gases.
   b. Liquids are viscous.
   c. There are forces of attraction between liquid particles but not between gas particles.
   d. There is no boundary surface in a gas.
   e. Molecules leave the liquid surface and become gaseous when they have sufficient velocity.

7. Most metal structures can be considered to be an arrangement of equal spheres packed together as closely as possible. Which of the following statements are true?

   a. In any close packed layer each sphere is in contact with six others.
   b. The second layer fits in the depressions of the first so that each sphere is in contact with three spheres in the lower layer.
   c. If a third layer is repeated over the first then we have a hexagonal close packed system (ABABAB system).
   d. If the third layer is laid so as not to be above the first (an ABCABC system) we have a cubic closed packed system or face centred lattice.
   e. In both systems each sphere is in contact with twelve others.

8. The diagram overleaf shows the limiting position for 6:1 co-ordination in an ionic crystal (i.e. where six ions of like charge are just touching, with an ion of opposite charge in the resulting space).
   Which of the following statements are true?

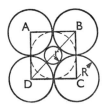

a. The diagonal of the square BD is $\sqrt{2}AB$.
b. Radius of the small sphere (r) is $0\cdot41 \times$ radius of the big sphere (R).
c. This type of co-ordination will not occur unless r/R is greater than $0\cdot41$.
d. This type of co-ordination will not occur unless r/R is less than $0\cdot41$.
e. Caesium chloride is an example of this type of packing.

9. The diagram shows the limiting position for 8:1 co-ordination (i.e. where eight ions of like charge are just touching an ion of opposite charge).
Which of the following statements are true?

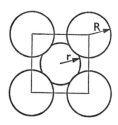

a. The eight larger ions are at the corners of a cube.
b. This is sometimes called a body centred system.
c. The radius of the small sphere is $(\sqrt{3} - 1) \times$ radius of the large sphere.
d. This type of co-ordination will not occur unless r/R is less than $0\cdot73$.
e. Sodium chloride is an example of this type of packing.

10. *Ionic radii in pm*
    Cs⁺ 168
    Rb⁺ 149
    Br⁻ 196
    I⁻   216
Which of the following statements are true?

a. In the crystal of RbI
$$\frac{r_{Rb}}{R_I} = 0\cdot69$$
b. In the crystal of CsBr
$$\frac{r_{Cs}}{R_{Br}} = 0\cdot85$$
c. We would expect rubidium iodide to be eight co-ordinated.

d. We would expect each caesium ion to be surrounded by six bromide ions.
e. We would expect caesium bromide to have the same lattice structure as caesium chloride.

11. Given that a mole of NaCl is 58·5 g, the distance between centres of the Na⁺ and Cl⁻ ions is 280 pm, and the density of NaCl is $2·16 \times 10^3$ kg m⁻³, find the wrong step in the following calculation of the Avogadro constant. (L, the number of ion pairs in a mole of NaCl.)

a. Volume of the unit cell of NaCl is $(2·8 \times 10^{-10})^3$ m³.
b. The unit cell consists of 4Na⁺ and 4Cl⁻ ions.
c. Each of these ions is shared between eight unit cells so the unit cell contains $\frac{1}{2}$Na⁺ and $\frac{1}{2}$Cl⁻ ions.
d. The mass of an ion pair is
$$\frac{58·5 \times 10^{-3} \text{ kg mol}^{-1}}{L \text{ mol}^{-1}}$$
e. $L = \dfrac{58·5 \times 10^{24}}{2·16 \times (2·8)^3}$ mol⁻¹

12. Covalent bonds are formed between shared pairs of electrons of opposing spin. If we make the simple assumption that the negatively charged clouds so formed will repel and orientate themselves as far away as possible from each other (whether the charge clouds contain a pair of bonded electrons or a lone pair), which of the following shapes of molecules would you expect?

a. $BeCl_2$ is linear.
b. H_2O is linear.
c. BF_3 is trigonal.
d. NH_3 is trigonal.
e. CH_4 is tetrahedral.

13. Berthollide compounds do not appear to agree with the Law of Definite Proportions (Constant Composition). Such compounds are ferrous sulphide and ferrous oxide. Which of the following seem to be reasonable statements?

a. An ideal ferrous oxide crystal would contain equal numbers of Fe^{2+} and O^{2-} ions.
b. If some of the iron was present as Fe^{3+} there would have to be a deficiency of O^{2-} ions.
c. Such a crystal as in b. would be expected to be a semi-conductor.
d. The actual composition of the compound in b. varies from $FeO_{0·86}$ to $FeO_{0·92}$.
e. Berthollide compounds can also occur due to a *deficiency* of anions (e.g. ZnO loses oxygen when hot and glows yellow).

STATES OF MATTER

Comparative values at 0°C for use in Questions 14–16.

	dipole moment	molar mass	dielectric constant	density	surface tension	viscosity	boiling point °C
Water HOH	1·84	18	81	1·00	74	1·0	100
Ethanol C_2H_5OH	1·85	46	25	0·79	22	1·2	78·5
Diethyl ether $C_2H_5OC_2H_5$	1·10	74	4·3	0·71	17	0·23	34·5
Benzene C_6H_6	0·00	78	2·3	0·88	29	0·65	80
Phenol C_6H_5OH	1·70	94	73	1·07	41		182
Toluene $C_6H_5CH_3$	0·37	92		0·87	28	0·59	110
Aniline $C_6H_5NH_2$	1·55	93	7·3	1·02	43	4·4	184
Nitrobenzene $C_6H_5NO_2$	4·23	123		1·20	44	2·0	211

14. Benzene, ethanol, and nitrobenzene were allowed to run from three similar burettes. A charged polythene rod was placed so as to be exactly 5 cm from each of the three jets. It was found that the benzene jet was unaffected, the ethanol jet was deflected 5° from the vertical, and the nitrobenzene jet was deflected 12°.
Which statement do you think provides the best hypothesis for further investigation?

a. Nitrobenzene has the highest surface tension.
b. Nitrobenzene has the highest viscosity.
c. Nitrobenzene has the highest relative molar mass.
d. Nitrobenzene has the highest dielectric constant.
e. Nitrobenzene jet has the greatest momentum.

15. Examination of the table of physical properties should show that there is a relationship between molecular size and structure, and physical properties.
Attraction between molecules within a liquid can result in a high boiling point, high surface tension, and dielectric constant.
Which statements do you think are true?

   a. We would not expect the molecules of water to have such a large attraction for each other from the figure quoted for the dipole moment of water.
   b. The high boiling point, surface tension, and dielectric constant are explained in terms of a special hydrogen bond in which hydrogen atoms link together two other atoms (usually O, N or F).
   c. The hydrogen bonding in ethanol is probably much more than in phenol.
   d. There is evidence for more hydrogen bonding in diethyl ether than in ethanol.
   e. There is considerable evidence of hydrogen bonding in nitrobenzene.

16. Which of the following estimates of properties do you think are reasonable?

   a. The dipole moment of phenyl ethyl ether ($C_6H_5OC_2H_5$) is about 4·5.
   b. The dipole moment of nitroethane ($C_2H_5NO_2$) is about 3·9.
   c. The boiling point of ethylbenzene ($C_6H_5C_2H_5$) is about 95°C.
   d. The boiling point of ethylamine ($C_2H_5NH_2$) is likely to be much less than that of ethanol (C_2H_5OH).
   e. We would expect the dielectric constant of toluene to be higher than that of benzene.

17. A colloidal solution of sulphur in water is known as a sol.
Which of the following statements are also true about such a sol?

   a. Sulphur is the disperse medium.
   b. The sol is hydrophilic.
   c. The colloidal sulphur can be precipitated by boiling.
   d. The colloidal sulphur can be precipitated by the addition of Al^{3+} ions.
   e. The sulphur particles are between 10^{-7} and 10^{-9} m in diameter.

18. Which of the following are true statements about colloids?

   a. Starch is a lyophobic sol.
   b. Many lyophilic sols form gels. A gel is an intermediate stage in the coagulation of a sol.
   c. The process of purifying a colloidal solution is called dialysis and can be carried out through parchment or cellophane.
   d. Lyophobic sols are irreversible.
   e. Colloidal gold is positively charged.

19. Which of the following can be considered most likely to be examples of the precipitation of lyophilic sols by electrolytes?

a. Aluminium sulphate is used to clarify water.
b. When measuring the pH of soils it is usual to shake up the soil with barium sulphate in water before adding the B.D.H. indicator.
c. Clay is precipitated at the mouth of rivers by the sea water.
d. Blood can be clotted by the use of a styptic pencil (potash alum).
e. White of egg is coagulated by boiling.

20. The number of millimoles of sodium chloride and magnesium chloride required to coagulate a litre each of two sols X and Y are as follows:

	X	Y
NaCl	51	46
$MgCl_2$	0·71	23

Which of the following would seem reasonable values to expect for the amount of other salts required to cause precipitation?

a. X will be precipitated by 0·09 millimoles of $AlCl_3$.
b. Y will be precipitated by 0·3 millimoles of $AlCl_3$.
c. X will be precipitated by 25 millimoles of Na_2SO_4.
d. Y will be precipitated by 0·6 millimoles of $MgSO_4$.
e. Y will be precipitated by 0·08 millimoles of NaH_2PO_4.

REACTION KINETICS

1. Which of the following factors will increase the rate of production of hydrogen from dilute sulphuric acid and zinc?

    a. Increase the acid concentration.
    b. Use concentrated acid instead of dilute.
    c. Add more zinc.
    d. Use powdered zinc.
    e. Warm the solution.

2. Pieces of $CaCO_3$ of standard size were added in excess to 20 ml M HCl and the volume of CO_2 liberated was measured every 10 seconds. A graph was drawn as (1) below. Variations in the size of the pieces of $CaCO_3$ and the concentration of acid gave the curves (2) (3) and (4).

    a. (2) refers to standard size pieces and 10 ml 2M HCl.
    b. (2) refers to standard size pieces and 5 ml 5M HCl.
    c. (4) refers to standard size pieces and 20 ml 0·5 M HCl.
    d. (3) refers to smaller pieces and 20 ml M HCl.
    e. From the equation one would expect to collect about 480 cm³ of CO_2 at room temperature and atmospheric pressure in experiment (1).

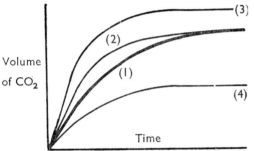

Which of the following statements are true?

3. The rate of a chemical reaction may be measured in a number of ways. Which of the following would be theoretically suitable for measuring the rate of decomposition of H_2O_2 in the presence of a platinum catalyst?

$$H_2O_2 \rightarrow H_2O + \tfrac{1}{2}O_2$$

    a. Mass of water produced at 10 second intervals.
    b. Total mass of H_2O_2, H_2O and O_2 at regular time intervals.
    c. Change in concentration of H_2O_2 each second.
    d. Change in refractive index of the liquid phase with time.
    e. Total volume of gas produced.

REACTION KINETICS

4. In the last question some of the methods were not very suitable in practice. Which of the following would be practical methods of measuring the rate of decomposition of hydrogen peroxide?

    a. Extract samples of the liquid and titrate with acid $KMnO_4$ at definite time intervals.
    b. Measure the increase in the volume of water.
    c. Measure the total volume of oxygen at regular time intervals.
    d. Measure the change in refractive of the liquid with time.
    e. Find the loss in weight every 10 seconds.

5. What changes in conditions are likely to increase the rate of decomposition of hydrogen peroxide significantly?

    a. An increase in temperature of 10 K.
    b. An increase in pressure of 10%.
    c. Addition of a homogeneous catalyst.
    d. Doubling the concentration of the hydrogen peroxide.
    e. Removal of the oxygen as fast as it is formed.

6. The catalytic decomposition of hydrogen peroxide by means of manganese (IV) oxide is said to be a first order reaction. It follows from this that:

    a. The stoichiometric equation should not be written as $2H_2O_2 \rightarrow 2H_2O + O_2$
    b. $\dfrac{d[H_2O_2]}{dt} = -k\,[H_2O_2]$
    c. $[H_2O_2]$ plotted against $\dfrac{1}{t}$ is a straight line
    d. $\log\,[H_2O_2]$ plotted against t is a straight line
    e. the time taken for half the hydrogen peroxide to decompose is independent of the original concentration of the hydrogen peroxide.

REACTION KINETICS

7. Butan-1-ol reacts with a conc. HBr/H_2SO_4 mixture to give 1-bromobutane.
$$C_4H_9OH + HBr \rightarrow C_4H_9Br + H_2O$$
The alcohol is miscible with the acid mixture but the bromobutane floats on top of the denser layer. If the thickness of the upper layer is measured in a vertical test tube at regular time intervals the graph is as shown:

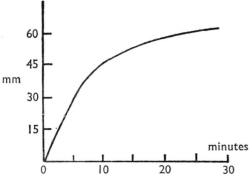

Which of the following conclusions appear to be reasonable?

a. It can be assumed that the rate of formation of bromobutane is the same as the rate of consumption of butanol.
b. The rate of the reaction decreases so that the amount of bromobutane is doubled every 5 minutes.
c. After 5 minutes half the total amount of bromobutane has been formed.
d. The reaction has a half life of 5 minutes.
e. The reaction appears to be first order with regard to butanol.

8. When the reaction in question 7 is repeated and the concentrations of HBr and H_2SO_4 are altered, neither is found to affect the reaction rate. Which mechanism for the rate determining step fits the experimental results?

a. $BuOH + HBr \rightarrow BuBr + H_2O$
b. $BuOH + H^+ \rightarrow BuOH_2^+$
c. $BuOH \rightarrow Bu^+ + OH^-$
d. $HBr \rightarrow H^+ + Br^-$
e. $BuOH + H_2SO_4 \rightarrow Bu^+ + H_2O + HSO_4^-$

9. Although the experiment of question 7 gave the results shown it is not an ideal method of studying reaction kinetics. Which of the following comments do you think are relevant?

a. Some loss of HBr could occur due to oxidation by conc. H_2SO_4.
b. Since one of the products is water it would dilute the lower layer.
c. It would be better to measure the rate of change of BuOH.
d. It would be better if the reaction took place under homogeneous conditions.
e. As HBr is used up the concentration of acid will alter and slow up the reaction.

REACTION KINETICS

Iodine reacts with acetone in dilute aqeous solution so that the system is homogeneous throughout. The stoichiometric equation is:

$$CH_3COCH_3 + I_2 \rightarrow CH_3COCH_2I + HI$$

A series of experiments were performed, the rate of reaction being determined by the decrease in iodine concentration with time as measured by a colorimeter, and the following results obtained:

Expt	[acetone]	[H$^+$]
1	1·0 M	0·10 M
2	1·0 M	0·20 M
3	2·0 M	0·10 M

10. a. In expt (1) the rate of decrease in iodine concentration is 2×10^{-4} mol l^{-1} min^{-1}
    b. Rate (3) is double rate (1)
    c. Rate (1) is doubled if the initial acetone concentration is doubled.
    d. Rate = k [acetone]
    e. Reaction is second order with respect to acetone.

11. a. Rate (2) is the same as rate (3)
    b. Rate (1) is doubled if the initial acid concentration is doubled.
    c. $\dfrac{d[I_2]}{dt} = k[H^+]$
    d. The slow step of the reaction must involve H$^+$ ions
    e. The reaction is first order with respect to hydrogen ions.

12. a. [I$_2$] decreases with time
    b. The reaction is zero order with respect to iodine.
    c. Iodine does not take part in the reaction.
    d. Iodine does not take part in the rate determining step.
    e. The reaction is second order overall.

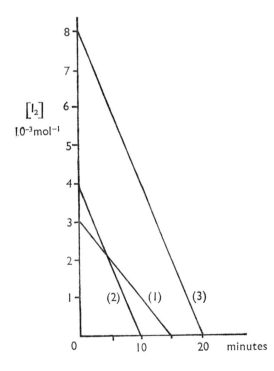

Which of the statements in questions 10–12 do you agree with?

REACTION KINETICS

13. In the reaction considered in the questions 10–12 which of the following hypotheses best fits the facts as a reasonable explanation of what takes place in the rate determining step?

a. $CH_3COCH_3 + I_2 \rightarrow CH_3COCH_2I + H^+ + I^-$
b. $CH_3COCH_3 + H^+ + I_2 \rightarrow CH_3COCH_2I + HI$
c. $CH_3COCH_3 \rightarrow CH_3 - \underset{\underset{OH}{|}}{C} = CH_2$
d. $CH_3COCH_3 + H^+ \rightarrow CH_3 - \underset{\underset{OH}{|}}{C^+} - CH_3$
e. $CH_3COCH_3 + I^+ \rightarrow CH_3COCH_2I + H^+$

14. Methyl acetate was saponified by treating it at constant temperature with an aqueous solution of sodium hydroxide. At the start of the experiment the concentrations of methyl acetate and sodium hydroxide were the same. Which of the following statements by themselves provide sufficient justification for saying that the reaction is second order overall?

a. One mole of CH_3COOCH_3 reacts with 1 mole of NaOH
b. Rate $= k$ [ester] [OH$^-$]
c. The half life of the ester is independent of the initial concentration of alkali.
d. The plot of log [ester] against time is a straight line
e. The plot of $\dfrac{1}{[\text{ester}]}$ against time is a straight line.

15. (A) If 2-iodo-2-methyl propane is shaken with aqueous AgNO$_3$ a yellow precipitate of AgI is seen after a few minutes.

(B) If 1-Iodobutane is shaken with aqueous AgNO$_3$ no precipitate is formed unless the iodobutane is shaken first with aqueous NaOH before shaking with the AgNO$_3$. If, however, this is done there is a copious precipitate of AgI. Which of the following are reasonable inferences?

a. Hydrolysis takes place in both reactions.
b. The reactions proceed by different mechanisms
c. (B) involves OH$^-$ ions in the rate determining step.
d. (A) does not require OH$^-$ ions in the rate determining step.
e. (A) could be first order.

80

REACTION KINETICS

16. Bromate ions react with bromide ions in acid solution according to the equation:
$BrO_3^- + 5Br^- + 6H^+ \rightarrow 3Br_2 + 3H_2O$.
The following results were obtained in four experiments with the initial quantities shown:

a. The reaction is first order with respect to BrO_3^- ions.
b. The reaction is first order with respect to Br^- ions.
c. The reaction is second order with respect to H^+ ions.
d. The reaction is third order overall.
e. $\dfrac{d[Br_2]}{dt} = k [H^+] [Br^-] [BrO_3^-]$

Expt.	ml M BrO_3^-	ml M Br^-	ml M H^+	total volume	relative rate
1	50	250	300	1 litre	1
2	50	250	600	1 litre	4
3	50	125	600	1 litre	2
4	100	250	600	1 litre	8

Which are valid inferences?

17. The graph below shows the comparison between the uncatalysed reaction between bromoethane and water and the same reaction when catalysed by iodide ions.

a. Heat of the uncatalysed reaction is E_1 joules given out.
b. Less heat is given out in the uncatalysed reaction.
c. E_2 is the uncatalysed activation energy.
d. E_3 is the catalysed activation energy.
e. X illustrates a possible intermediate compound.

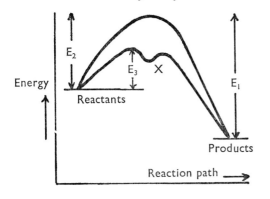

Which statements are true?

REACTION KINETICS

18. The distribution of molecular velocities is shown for a gaseous reaction.

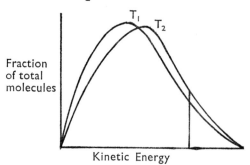

It is found that raising the temperature by 10 K doubles the rate of the reaction although the average velocity of the molecules is only increased slightly. Which appear to be logical conclusions?

a. There is twice the chance of collisions between molecules.
b. The number of effective collisions between molecules is doubled.
c. The proportion of molecules with higher velocity is now much greater.
d. The activation energy of the reaction is less at the higher temperature.
e. Each collision now leads to reaction.

19. Radium decays according to the reaction
$$^{226}_{88}Ra \rightarrow {}^{222}_{86}Rn + {}^{4}_{2}He$$
and is said to obey the decay law
$$N = N_0 \, e^{-kt}$$
where N is the number of atoms of radium at time t and N_0 is the number of atoms at an arbitary zero of time. Which of the following statements are true?

a. The time of decay is directly proportional to the number of atoms present.
b. $\dfrac{dN}{dt} = -kN$
c. This is a first order reaction.
d. $\log_e \dfrac{N}{N_0} = -k\,t$
e. The time taken for half the radium to decay is independent of the number of atoms present.

20. The radioactivity of living material due to $^{14}_{6}C$ is 15 counts each minute for each gram of carbon. On the death of the plant the $^{14}_{6}C$ decays emitting β rays so that the half life is 5600 years.
A piece of linen from an Egyptian tomb was found to show 10 counts each minute for each gram of carbon present. Which statements are true?

a. $\tfrac{2}{3}$ of the original $^{14}_{6}C$ remains.
b. The linen was $\tfrac{2}{3} \times 5600$ years old.
c. All the $^{14}_{6}C$ will have decayed to $^{14}_{7}N$ in 11 200 years
d. $k = \dfrac{\log_e 2}{5600}$
e. The age is given by $\dfrac{\log 1.5}{\log 2} \times 5600$ years.

PERIODIC TABLE

1. Atomic number can be defined as
   a. the order of elements when arranged according to their atomic masses.
   b. the number of protons in the element.
   c. the number of electrons in an unionised atom.
   d. the number of protons in an atom.
   e. the number of neutrons in an atom.

2. Certain pairs of elements (e.g. Te and I) appear to be in the wrong order in the modern periodic table when arranged according to their atomic masses. Which statements are correct concerning this?
   a. The atomic masses are wrong.
   b. The order according to mass numbers is correct.
   c. The order according to atomic numbers is correct.
   d. The elements are so similar chemically that order does not matter.
   e. The order arises because of the unusual isotopic composition of one of the elements.

3. Which statements are correct concerning the radii of atoms and ions?
   a. Na^+ has a smaller radius than Na.
   b. Ra has larger radius than Be.
   c. Sr has a smaller radius than Rb.
   d. Cl has a smaller van der Waals radius than F.
   e. C has a smaller covalent radius than Si.

4. Which oxides are acidic, or more acidic than basic?
   a. Li_2O
   b. CaO
   c. B_2O_3
   d. CO_2
   e. N_2O_4

5. Which chlorides have all of these properties?
   i. Low melting and boiling points.
   ii. Are non-electrolytes when fused.
   iii. Are hydrolysed by water.
   a. RbCl
   b. $SrCl_2$
   c. CCl_4
   d. PCl_3
   e. PCl_5

6. Which hydrides have high melting points and are also good electrolytes when fused?
   a. NaH
   b. $(AlH_3)_n$
   c. SiH_4
   d. H_2S
   e. HCl

7. Which statements are correct concerning changes occurring as the relevant group is descended?

   a. Caesium reacts more vigorously with oxygen, water, and halogens than does sodium.
   b. Ba ionises to Ba^{2+} more readily than Mg to Mg^{2+} when placed in water.
   c. Carbon is a non-metal, whereas lead is essentially metallic.
   d. A bismuth sulphate exists, but not a nitrogen sulphate.
   e. Fluorine is a better oxidising agent than iodine.

8. Which statements are correct concerning changes occurring as group $2M$ is descended?

   a. The atomic radii increase.
   b. The atomic volumes increase.
   c. The electronegativities increase.
   d. The first ionisation potentials decrease.
   e. The ionic radii decrease.

9. What are the properties you would expect of an element in group $6T$?

   a. It forms XO_4^{2-}.
   b. It forms XO_3.
   c. It has various valencies including 2 and 6.
   d. Its compounds are often catalysts.
   e. It forms many coloured salts.

10. Which statements are correct concerning the so-called 'inert' gases?

    a. There is no attraction at all between the atoms in the gas.
    b. A sample of nitrogen contaminated with Kr and Xe is less dense than pure nitrogen.
    c. When radioactive isotopes decay by alpha particle emission, He is produced.
    d. They form no chemical compounds.
    e. He is sometimes used with oxygen in deep sea diving.

11. *Electron energy levels in different elements*

	1s	2s	2p	3s	3p	3d	4s	4p	4d	5s	5p
A	2	2	3								
D	2	2	6	2	2						
E	2	2	6	2	6	3	2				
G	2	2	6	2	6	10	2	6		2	
J	2	2	6	2	6	10	2	6	10	2	5

    Property
    a. A^{3+} occurs in ACl_3.
    b. D^{4+} occurs in DI_4.
    c. E is a transition metal.
    d. G^{2+} occurs in GCl_2.
    e. J^- occurs in KJ.

    In which cases does the property stated agree with what you would expect from the electron arrangement?

12. Which atoms and ions have the same electron arrangement?
   a. S^{2-}
   b. Ar
   c. Sr^{2+}
   d. K^+
   e. Cl^-

13. In a comprehensive periodic table, fundamental information concerning each element is given as shown:

20	Symbol	40·08
	$M \to M^+$	590 kJ mol^{-1}
	$M^+ \to M^{2+}$	1150 kJ mol^{-1}
197 pm	$(Ar)(4s)^2$	99 pm

   Which statements MUST be correct?
   a. The element is strontium.
   b. The electron structure of the atom is 2 8 8 2.
   c. The ionisation energy for $M \to M^{2+}$ is 1740 kJ mol^{-1} evolved.
   d. The radius of M^{2+} is 197 pm.
   e. There are isotopes of mass numbers 40 and 41.

14. In which of these would you expect the fluorine or hydrogen atoms to be arranged tetrahedrally in space around the other atom so that the bond angle is approximately 109°?
   a. BF_3
   b. BF_4^-
   c. CH_4
   d. NH_3
   e. H_2O

15. Which statements are correct concerning these two isotopes?
   $^{40}_{18}Ar$ and $^{40}_{20}Ca$
   a. Ar has 18 protons and 40 neutrons per atom in this isotope.
   b. The atomic mass of naturally occurring calcium must be 40·0.
   c. Both isotopes have 40 electrons per atom.
   d. Both isotopes have mass numbers of 40.
   e. They are isobars.

16. Copper has only two natural isotopes:
   $^{63}_{29}Cu$ and $^{65}_{29}Cu$
   The atomic mass is 63·54.
   What is the percentage by mass of the lighter isotope?
   a. 65 per cent
   b. 73 per cent
   c. 79 per cent
   d. 84 per cent
   e. 90 per cent

17. The following were among the first nuclear reactions to be detected.
   Which transformations are correctly stated?
   a. $^{14}_{7}N + ^{4}_{2}He \to ^{1}_{1}H + ^{17}_{8}O$ (Rutherford, 1919)
   b. $^{7}_{3}Li + ^{1}_{1}H \to ^{4}_{2}He + ^{4}_{2}He$ (Cockroft and Walton, 1932)
   c. $^{27}_{13}Al + ^{4}_{2}He \to ^{30}_{15}P + ^{1}_{0}n$ (Curie-Joliot, 1933)
   d. Be bombarded with alpha particles gave S. (Curie-Joliot, 1933)
   e. Mg bombarded with alpha particles gave Si. (Curie-Joliot, 1933)

18. $^{226}_{88}$Ra decays by alpha particle emission. Which isotope would you expect to be produced?

    a. $^{222}_{86}$Rn
    b. $^{224}_{86}$Rn
    c. $^{222}_{88}$Ra
    d. $^{222}_{90}$Tl
    e. $^{228}_{90}$Th

19. Which isotopes are radioactive?

    a. $^{90}_{38}$Sr
    b. $^{131}_{53}$I
    c. $^{3}_{1}$H
    d. $^{14}_{6}$C
    e. $^{4}_{2}$He

20. A recent attempt has been made to synthesize element 104. Which method is most likely to succeed (in theory)?

    a. Bombard U with protons.
    b. Oxidize element 103 (Lw).
    c. Bombard element 98 (Cf) with neutrons.
    d. Bombard Bk with $^{13}_{6}$C nucleii.
    e. Bombard Pu with neon nucleii.

GROUP 1 *M* THE ALKALI METALS

1. The metallic radii of the alkali metals (except francium) are given in **a.** to **e.** Deduce the value for rubidium.
   - **a.** 235 pm
   - **b.** 216 pm
   - **c.** 203 pm
   - **d.** 157 pm
   - **e.** 123 pm

2. The ionisation energy for sodium is $Na(g) \rightarrow Na^+(g)$: $\Delta H = +494$ kJ mol^{-1}. The corresponding value for potassium will be
   - **a.** $+800$ kJ mol^{-1}
   - **b.** $+519$ kJ mol^{-1}
   - **c.** $+494$ kJ mol^{-1}
   - **d.** $+418$ kJ mol^{-1}
   - **e.** -160 kJ mol^{-1}

3. Indicate which electron structures are those of the alkali metals.
   - **a.** 2 8 18 18 8 1
   - **b.** 2 8 18 32 18 8 1
   - **c.** 2 1
   - **d.** 2 8 18 1
   - **e.** 2 8 18 32 18 1

4. When placed in cold water so that it can move freely, an element reacts so violently that the gas liberated is ignited. The element could be:
   - **a.** Li
   - **b.** Ca
   - **c.** Cs
   - **d.** Na
   - **e.** K

5. Which of the alkali metals has the smallest ratio $\dfrac{\text{Atomic mass}}{\text{(Density at 20°C)}}$?
   - **a.** Li
   - **b.** Na
   - **c.** K
   - **d.** Rb
   - **e.** Cs

6. Sodium is used industrially in the production of
   - **a.** lead tetraethyl antiknock for petrol
   - **b.** sodium peroxide
   - **c.** sodium cyanide
   - **d.** sodium hydroxide
   - **e.** titanium

7. Sodium was heated in air until none remained unreacted. 0·71 g of the product was added to cold excess acidified potassium iodide solution. To use up the

liberated iodine, 20 ml of 0·1 M sodium thiosulphate were needed. Calculate the percentage of sodium peroxide in the sample (to the nearest 1 per cent).
[$I_2(s) + 2Na_2S_2O_3(aq) \rightarrow Na_2S_4O_6(aq) + 2NaI(aq)$]

a. 2 per cent
b. 6 per cent
c. 11 per cent
d. 16 per cent
e. 21 per cent

8. When sodium is left to stand in air in this country for a long time the final product will be:

a. Na_2O
b. NaOH
c. Na_2CO_3
d. $Na_2CO_3, 10H_2O$
e. Na_2CO_3, H_2O

9. A compound which can be used in submarines both to absorb carbon dioxide and liberate oxygen is

a. NaOH
b. Na_2O
c. Na_2O_2
d. NaH
e. Soda lime

10. Indicate the correct statements concerning the hydrides of the alkali metals.

a. Lithium hydride is $H^+ Li^-$.
b. Sodium hydride is $Na^+ H^-$.
c. Rubidium hydride is covalent.
d. 1 mole of sodium hydride liberates 2 moles of hydrogen with excess water.
e. They can be made by oxidizing the metal with hydrogen.

11. A white crystalline solid decomposed readily on warming, giving a colourless gas. The remaining solid was dissolved in water. A 0·1 M solution had a pH greater than 10.
The original solid could have been

a. $NaHCO_3$
b. Na_2CO_3
c. Na_2SO_4
d. $RbHCO_3$
e. Cs_2CO_3

12. A white solid effervesced in dilute hydrochloric acid, and a 0·1 M solution in water had a pH of approximately 8.
The solid could have been

a. K_2CO_3
b. $KHCO_3$
c. K_2SO_4
d. $KHSO_4$
e. K_2SO_3

GROUP 1 M

13. A white crystalline solid is strongly acidic in dilute solution, and gives a thick white precipitate when excess dilute hydrochloric acid and barium chloride solution are added together.
It could be

   a. NaHSO₄
   b. Na₂SO₄
   c. NaHSO₃
   d. Na₂SO₃
   e. 'Harpic'

14. A white compound on strong heating liberates oxygen. The solid product is acidified with dilute hydrochloric acid, liberating brown fumes.
The compound could be

   a. NaNO₃
   b. NaNO₂
   c. NaClO₄
   d. NaBrO₃
   e. RbNO₃

15. Indicate the correct statements concerning the structures of the chlorides of sodium, caesium, and francium.

   a. The number of Cl⁻ ions touching each Na⁺ ion is six.
   b. The Cl⁻ ions are arranged octahedrally around the Na⁺ ions.
   c. The co-ordination number of the caesium ion is eight.
   d. The Cl⁻ ions are arranged tetrahedrally around the Cs⁺ ions.
   e. FrCl will probably be similar to NaCl in structure.

16. Indicate the properties you would expect francium to have.

   a. The hydroxide will be soluble in water.
   b. The metal will readily react with water.
   c. The carbonate will be alkaline in solution.
   d. The carbonate will decompose readily on warming.
   e. The hydrogen sulphate will be alkaline in solution.

17. An element has these properties:
   i. It is not readily tarnished in air.
   ii. The carbonate is readily decomposed by heat.
   iii. The chloride is hydrolysed in water.
The element is probably

   a. Li
   b. Na
   c. Rb
   d. K
   e. Cs

18. Indicate the statements which are correct concerning the Solvay (ammonia-soda) process.

   a. This equilibrium occurs in the tower
   $H_2O(l) + CO_2(g) \rightleftharpoons HCO_3^-(aq) + H^+(aq)$
   b. Sodium hydrogen carbonate precipitates in

the industrial process even though it is soluble in water.
c. A catalyst is used as the equilibrium is reached slowly.
d. The carbonation reaction is carried out at about 80°C.
e. Almost all the ammonia used in the reaction is regenerated from the ammonium chloride produced by heating with sodium hydroxide.

19. Indicate the correct statements concerning the industrial production and uses of sodium hydroxide and carbonate.

a. Sodium hydroxide is manufactured by the ammonia-soda process.
b. Sodium carbonate is manufactured by the lime-soda process.
c. Sodium hydroxide is manufactured in a Downs cell.
d. Sodium hydroxide is manufactured by electrolysis in a mercury cell.
e. Sodium hydroxide is used in the rayon industry.

20. Solubilities of various salts are given (in g per 100 g of water). Plot these carefully and deduce which statements are correct.

Temperature (°C)	10	20	30	40	50	60	70	80
Sodium chloride	35·8	36·0	36·3	36·6	37·0	37·3	37·8	38·4
Potassium chloride	31	34	37	40	43	46	48	51
Sodium nitrate	80	88	96	104	114	124		148
Potassium nitrate	21	32	46	64	86	110	138	169
Sodium sulphate	9	19	41	49	47	45		44

a. On mixing almost saturated solutions of sodium nitrate and potassium chloride at 80°C and cooling to 50°C, the first salt to precipitate as crystals will be potassium nitrate.
b. The solubility of sodium nitrate at 70°C is 141 g.
c. It is impracticable to prepare pure sodium chloride on an industrial scale by cooling a saturated solution without evaporation.
d. The sodium sulphate curve between 10°C and 30°C is the solubility curve of anhydrous sodium sulphate and that after 40°C is of hydrated sodium sulphate.
e. The temperature at which this equilibrium exists is 40°C.
$$Na_2SO_4 \rightleftharpoons Na_2SO_4, 10H_2O$$

GROUP 2 M

1. The ionic radii of group 2M elements (excluding Ra) are given in **a.** to **e.** Deduce the value for Sr.

   a. 135 pm
   b. 113 pm
   c. 99 pm
   d. 65 pm
   e. 31 pm

2. The ionisation energies for group 2M elements (excluding Ra) are given in **a.** to **e.** Deduce the value for Mg. i.e. $M(g) \rightarrow M^{2+}(g) + 2e^-$: $\Delta H = ?$ kJ mol^{-1}

   a. + 1468 kJ mol^{-1}
   b. + 1608 kJ mol^{-1}
   c. + 1740 kJ mol^{-1}
   d. + 2186 kJ mol^{-1}
   e. + 2660 kJ mol^{-1}

3. Standard redox potentials are given.
   $Cs^+(aq) + e^- \rightleftharpoons Cs(s)$
   $\quad E^\circ = -2.92$ volts
   $Ca^{2+}(aq) + 2e^- \rightleftharpoons Ca(s)$
   $\quad E^\circ = -2.87$ volts
   $2H^+(aq) + 2e^- \rightleftharpoons H_2(g)$
   $\quad E^\circ = 0.00$ volts
   $Cl_2(aq) + 2e^- \rightleftharpoons 2Cl^-(aq)$
   $\quad E^\circ = +1.40$ volts
   What value would you expect for $Ba^{2+}(aq) + 2e^- \rightleftharpoons Ba(s)$?

   a. −2.92 volts
   b. −2.91 volts
   c. −2.75 volts
   d. −1.45 volts
   e. +0.23 volts

4. The atomic radius of calcium is 197 pm and of potassium is 231 pm.
   By comparing calcium with potassium, which of the following statements would you expect to be correct?

   a. The ionisation energy for $K(s) \rightarrow K^+(g)$ will be greater than for $Ca(s) \rightarrow Ca^{2+}(g)$.
   b. The standard electrode potential of potassium will be more negative than that of calcium.
   c. The extent of hydration of Ca^{2+} in aqueous solution will be greater than that of K^+.
   d. Calcium will be more reactive towards water and air than potassium.
   e. As the calcium atom is the smaller there will be more atoms in a mole of calcium than in a mole of potassium.

5. 25·0 ml of lime water saturated at 18°C needed 11·5 ml 0·1 M HCl in a titration using screened methyl orange. 15 ml of

another sample saturated at 95°C needed 2·7 ml of 0·1 M HCl.
Which statements are correct?

a. The solubility of calcium hydroxide is 0·17 g per 100 g of water at 18°C.
b. The solubility is greater at 95°C.
c. In the equilibrium
   solid lime $\rightleftharpoons$ saturated solution
   the change from left to right is exothermic.
d. The first titration would be 23·0 ml if 0·1 M HNO_3 were to be used instead.
e. The first titration would be as stated if phenolphthalein were to be used as the indicator.

6. Calculate which source would give the largest weight of magnesium if it could all be extracted.

a. 1 mole of Carnallite
b. 1 mole of Dolomite
c. 1 mole of Magnesite
d. 1 mole of Epsom salts
e. 25 litres of sea water containing 0·44 g of magnesium chloride per 100 ml.

7. 0·72 g of magnesium on heating in air gives 1·15 g of product. The product corresponds to

a. MgO
b. Mg_3N_2
c. MgO + Mg_3N_2
d. $MgCO_3$
e. MgO_2

8. Anhydrous magnesium chloride can be prepared by

a. evaporating magnesium chloride solution to dryness.
b. heating magnesium in dry chlorine.
c. heating magnesium in dry hydrogen chloride.
d. heating $MgCl_2,6H_2O$ in dry hydrogen chloride.
e. heating a mixture of magnesium oxide and carbon in dry chlorine.

9. Which will give the largest volume of hydrogen when 1 g is added to excess water?
(All volumes measured at the same temperature and pressure.)

a. Na
b. Ca
c. LiH
d. NaH
e. CaH_2

10. The chemical reactions represented by the equation
$$CaCO_3(s) + H_2O(l) + CO_2(g) \rightleftharpoons Ca(HCO_3)_2(aq)$$
are responsible for:
   a. hardness in water.
   b. scale in hot water pipes.
   c. the setting of plaster of Paris.
   d. the weathering of mortar.
   e. stalactite formation.

11. Carbonate hardness in water caused by calcium and magnesium salts can be largely, if not entirely, removed by
   a. boiling.
   b. adding washing soda.
   c. passing through an anion exchange column.
   d. adding sodium hexametaphosphate.
   e. adding ammonia solution.

12. In an industrial process anhydrite, coke, and certain oxides are heated together at 1200°C or more. This gives rise to:
   a. sulphur dioxide.
   b. cement.
   c. plaster of Paris.
   d. calcium carbide.
   e. calcium cyanamide.

13. A white solid shaken with water gives an alkaline solution. Carbon dioxide produces a white precipitate when bubbled into this dilute solution for a few moments. The solid could be:
   a. NaOH
   b. RbOH
   c. $Ca(OH)_2$
   d. $Sr(OH)_2$
   e. $Ba(OH)_2$

14. A white solid effervesces with dilute hydrochloric acid. The solution formed is then concentrated, and on adding dilute sulphuric acid a white precipitate appears. The original solid could be:
   a. $MgCO_3$
   b. $MgCO_3, Mg(OH)_2$
   c. $CaCO_3$
   d. $SrCO_3$
   e. $BaCO_3$

15. Two solids are mixed and added to water, when both dissolve giving a colourless solution.
   i. Addition of excess dilute sulphuric acid gives a white precipitate.
   ii. Addition of excess sodium hydroxide solution to the filtrate from i. gives a white precipitate.
   The mixture could be
   a. $BaCl_2 + Mg(NO_3)_2$
   b. $BaCl_2 + MgSO_4$
   c. $RaCl_2 + MgSO_4$
   d. $BaCl_2 + RaCl_2$
   e. $BaSO_4 + Sr(NO_3)_2$

GROUP 2 M

16. A dilute metal nitrate solution gives a white precipitate when dilute sulphuric acid is added, but no precipitate when dilute sodium hydroxide is added to separate samples.
The metal could be

   a. Li
   b. Cs
   c. Be
   d. Mg
   e. Ba

17. Addition of sodium carbonate solution to magnesium sulphate solution gives a basic carbonate.
The normal carbonate can be made by:

   a. adding sodium hydrogen carbonate solution to magnesium sulphate solution.
   b. adding potassium carbonate solution to magnesium sulphate solution.
   c. bubbling in carbon dioxide while slowly adding the sodium carbonate solution to magnesium sulphate solution.
   d. bubbling carbon dioxide into a solution of Epsom salts until there is no further reaction.
   e. boiling the suspension of basic carbonate.

18. Indicate the properties you would expect of radium and its compounds.

   a. $Ra(OH)_2$ will be soluble in water.
   b. $RaSO_4$ will be soluble in water.
   c. $RaCO_3$ will be readily decomposed by heating in a Bunsen burner.
   d. $RaCl_2$ will be hydrolysed in solution.
   e. Ra should react with water.

19. An element has these properties
   i. The hydroxide is soluble in sodium hydroxide solution.
   ii. The chloride is a poor electrical conductor when fused.
   iii. It does not react at all readily with water.
The element could be:

   a. Be
   b. Mg
   c. Sr
   d. Ba
   e. Ra

20. Two elements have similar properties.
   i. Their chlorides are both hydrolysed in water and are deliquescent.
   ii. They form carbonates which readily decompose on heating.
   iii. They form nitrides.
The elements could be:

   a. Li and Mg
   b. Na and Mg
   c. Na and Ca
   d. Be and B
   e. B and Si

GROUP 3 M

1. The ionisation energy for aluminium is:
   $Al_{(g)} \rightarrow Al^{3+}{}_{(g)} + 3e^- : \Delta H$
   $= +5137$ kJ mol^{-1}
   The value in the same units for
   $B_{(g)} \rightarrow B^{3+}{}_{(g)} + 3e^-$
   will probably be

   a. -4130
   b. -4340
   c. $+4762$
   d. $+5137$
   e. $+6879$

2. Standard redox potentials for three elements in the same period are given below:
   $Na^+{}_{(aq)} + e^- \rightleftharpoons Na_{(s)}$:
   $\quad E^\circ = -2\cdot71$ volts
   $Mg^{2+}{}_{(aq)} + 2e^- \rightleftharpoons Mg_{(s)}$:
   $\quad E^\circ = -2\cdot37$ volts
   $Cl_{2(aq)} + 2e^- \rightleftharpoons 2Cl^-{}_{(aq)}$:
   $\quad E^\circ = +1\cdot40$ volts
   What value would you expect for
   $Al^{3+}{}_{(aq)} + 3e^- \rightleftharpoons Al_{(s)}$?

   a. $-2\cdot84$ volts
   b. $-2\cdot50$ volts
   c. $-1\cdot66$ volts
   d. $+0\cdot14$ volts
   e. $+1\cdot39$ volts

3. An element has these properties:
   i. It never forms positive ions in chemical reactions.
   ii. The oxide is acidic.
   iii. The chloride is covalent and readily hydrolysed by water.
   It could be:

   a. B
   b. Al
   c. Ga
   d. In
   e. Tl

4. A chloride combines with ammonia to form a compound. One molecule of this contains one molecule of ammonia and one of the chloride joined together. The chloride could be:

   a. NaCl
   b. RaCl$_2$
   c. BCl$_3$
   d. AlCl$_3$
   e. CCl$_4$

5. A compound such as that formed in Question 4 arises because:

   a. ammonia is an electron donor.
   b. the chloride is an electron acceptor.
   c. ammonia is an electron acceptor.
   d. the chloride is an electron donor.
   e. the chloride is electron deficient.

6. These are *facts* given in a well-known textbook published in 1950. Which facts agree with the changes you would expect as a group is descended?

   a. The alkaline properties of $Ga(OH)_3$ are much weaker than those of $Al(OH)_3$.
   b. $Ga(C_2H_5)_3$ is spontaneously flammable in air.
   c. Indium oxide can be made by heating indium carbonate.
   d. $Tl(OH)_3$ does not exist.
   e. Thallium nitrate is decomposed by heat to Tl_2O_3 and oxides of nitrogen.

7. The heats of certain reactions are given in kJ:
   $$4Al(s) + 3O_2(g) \rightarrow 2Al_2O_3(s) : \Delta H = -3260$$
   Calculate the heat change per mole of oxygen involved, and thus deduce which oxides are likely to be reduced to the element by aluminium.

   ΔH
   a. $2Fe(s) + 3O_2(g) \rightarrow 2Fe_2O_3(s)$: -1588
   b. $2Mg(s) + O_2(g) \rightarrow 2MgO(s)$: -1213
   c. $Si(s) + O_2(g) \rightarrow SiO_2(s)$: -8280
   d. $2Cr(s) + 3O_2(g) \rightarrow 2Cr_2O_3(s)$: -2230
   e. $Ti(s) + O_2(g) \rightarrow TiO_2(s)$: -1800

8. Indicate the reagents which rapidly attack and destroy a pure aluminium saucepan lid.

   a. Oxygen at 25°C.
   b. Conc. nitric acid.
   c. Conc. hydrochloric acid.
   d. Hot caustic soda solution.
   e. Hot washing soda solution.

9. Which reagents are likely to attack and destroy an anodised aluminium saucepan lid?

   a. Oxygen at 25°C.
   b. Conc. nitric acid.
   c. Conc. hydrochloric acid.
   d. Hot caustic soda solution.
   e. Hot washing soda solution.

10. Which of these statements concerning "alums" are correct?

    a. Potassium hydroxide solution is warmed with aluminium until no more reacts. After removing the excess aluminium, the filtrate is acidified with dilute sulphuric acid. On crystallising, aluminium potassium sulphate results.
    b. 0·1 mole of Iron(II) sulphate is completely oxidized with hydrogen peroxide in sulphuric acid. To this is added 0·1 mole of ammonium sulphate in water. On crystal-

lising, ammonium iron(III) sulphate results with no excess iron or ammonium salts left in solution.
c. Aluminium potassium sulphate and ammonium iron(III) sulphate are isomorphs.
d. 0·1 mole of potassium dichromate in water is completely reduced with a stream of sulphur dioxide. Crystallisation gives only chromium(III) potassium sulphate, with no excess potassium or chromium salts left in solution.
e. Chromium(III) potassium sulphate is a complex compound.

11. Which of these compounds would you expect to exist and have the same crystal shape as aluminium potassium sulphate?

a. Ammonium gallium alum.
b. Iron(II) ammonium sulphate.
c. Caesium indium sulphate.
d. $NaB(SO_4)_2, 12H_2O$.
e. Rubidium chromium sulphate.

12. A scheme to separate three hydrated oxides is as follows:
The mixture is warmed with excess sodium hydroxide solution. One oxide remains undissolved and is removed by filtration. Excess dilute hydrochloric acid is added slowly to the filtrate, when the second hydrated oxide precipitates. After filtering, sodium hydroxide solution is carefully added to the filtrate, precipitating the third hydrated oxide.
Which mixture of hydrated oxides could it have been?

a. $MgO + CaO + Al_2O_3$
b. $Al_2O_3 + SiO_2 + Fe_2O_3$
c. $Li_2O + B_2O_3 + SiO_2$
d. $BeO + Al_2O_3 + SiO_2$
e. $MgO + Al_2O_3 + SiO_2$

GROUP 3 M

13. Which statements concerning the extraction of aluminium are correct?
    a. Articles made from the metal were unknown before the introduction of the electrolytic method of extraction.
    b. It can be extracted by heating aluminium chloride with sodium.
    c. In the modern process, a high voltage and low current are used in the cell.
    d. Cryolite is used to lower the melting point of the electrolyte in c.
    e. The carbon cathode is rapidly oxidized in c.

14. A reagent, when added in excess to a solution containing copper and aluminium ions, precipitates only aluminium hydroxide.
    The reagent could be:
    a. $NaOH(aq)$
    b. $NH_3(aq)$
    c. $NaOH(aq) + NH_3(aq)$
    d. $NH_3(aq) + NH_4Cl(aq)$
    e. $NaOH(aq) + NH_4Cl(aq)$

15. Indicate the statements which you think are correct.
    a. When sodium carbonate solution is added to molar aluminium sulphate solution, aluminium carbonate is precipitated which immediately decomposes to aluminium hydroxide and carbon dioxide.
    b. Al_2S_3 cannot be prepared in solution because when formed it is instantly hydrolysed by water.
    c. Al^{3+} in water is probably associated with 6 H_2O.
    d. $Al^{3+}(aq)$ is a stronger acid than $H_2CO_3(aq)$ or $H_2S(aq)$.
    e. Aluminium sulphate and sodium hydrogen carbonate are used together in foam fire extinguishers.

16. By considering the relative sizes of and charges on the relevant ions, indicate which compounds are likely to form hydrated ions and an acidic solution when they are dissolved in water to give a molar solution.
    a. $RbCl$
    b. $Al(NO_3)_3$
    c. H_2SO_4
    d. $BeCl_2$
    e. Fr_2SO_4

17. Indicate those compounds which, on boiling an aqueous solution to dryness, will leave the anhydrous chloride.

a. CsCl
b. BeCl$_2$
c. MgCl$_2$
d. AlCl$_3$
e. FeCl$_3$

18. Which compounds are most likely to be completely ionic in the solid state?

a. CsF
b. BeI$_2$
c. AlBr$_3$
d. AlI$_3$
e. RbCl

19. Indicate the statements which you would expect to be correct concerning aluminium chloride under various conditions, knowing these properties:
  i. It sublimes at 178°C.
  ii. It is soluble in organic solvents.
  iii. The molar mass at 200°C is 267.

a. The solid consists of Al^{3+} 3Cl$^-$.
b. The solid and vapour have the structure

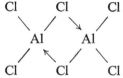

c. The solid will be readily hydrolysed by water.
d. In ether, it combines with the solvent to form the molecule [(C$_2$H$_5$)$_2$O → AlCl$_3$].
e. The solution in carbon tetrachloride conducts electricity.

20. In 1871 Mendeleeff predicted the properties of the then unknown element *eka*-aluminium ('below aluminium') from the known properties of elements near it in the periodic table.
Which of the following predictions concerning the element directly beneath aluminium appear reasonable?

a. It should have an equivalent of about 23 and an atomic weight of about 69.
b. It should form an alum with ammonium sulphate.
c. It might be extracted by reducing the trichloride with sodium.
d. Sodium carbonate added to a solution of the sulphate should give a white precipitate.
e. The element should liberate hydrogen with dilute hydrochloric acid and also with potassium hydroxide solution.

GROUP 4 M

1. Ionisation energies are given below in kJ mol⁻¹.
   $C(g) \rightarrow C^{2+}(g): \Delta H = +3440$
   $Si(g) \rightarrow Si^{2+}(g): \Delta H = +2340$
   $Sn(g) \rightarrow Sn^{2+}(g): \Delta H = +2092$
   $Pb(g) \rightarrow Pb^{2+}(g): \Delta H = +2150$
   Which deductions do you think are reasonable?

   a. The value for carbon is so high that the element will not form C^{2+} in compounds.
   b. Silicon is unlikely to form ionic compounds.
   c. The value of ΔH for $Ge(g) \rightarrow Ge^{2+}(g)$ will be + 2290 kJ mol⁻¹.
   d. The value of ΔH for $Sn(g) \rightarrow Sn^{4+}(g)$ will be +1820 kJ mol⁻¹.
   e. $Pb(g) \rightarrow Pb^{2+}_{(aq)}: \Delta H = +2150$ kJ mol⁻¹.

2.

	C	Si	Ge	Sn	Pb	
Atomic volume: (cm³ mol⁻¹)	(a) (Graphite)	2·33	5·36	7·31	11·34	a. 2·57
Melting point (°C)	3570 (Graphite)	(b)	958	232	327	b. 1414
First ionisation energy (kJ mol⁻¹)	1090	786	(c)	707	716	c. 870
Crystal radius for M^{4+}(pm)	15	41	53	(d)	84	d. 54
Electro-negativity (Pauling values)	2·5	1·8	1·7	1·7	(e)	e. 1·2

The table lists data from textbooks concerning these elements. In which cases do the values given in **a.** to **e.** match the expected values deduced from the table?

3. Which statements are correct?
   a. Sn forms salts less readily than Si.
   b. PbH_4 is more stable to heat than SiH_4.
   c. SnO is more basic than CO.
   d. Sn^{II} is more easily oxidized in aqueous solution than Sn^{IV}.
   e. Pb^{IV} is readily reduced to Pb^{II}.

4. A metal hydroxide dissolves in cold dilute hydrochloric acid. Careful addition of sodium hydroxide solution gives a white precipitate which dissolves as more is added. The hydroxide could be:
   a. $Be(OH)_2$
   b. $Mg(OH)_2$
   c. $Al(OH)_3$
   d. $Sn(OH)_2$
   e. $Pb(OH)_2$

5. An element forms a tetrachloride which is readily hydrolysed by water. It could be:

a. CCl_4
b. $SiCl_4$
c. $GeCl_4$
d. $SnCl_4$
e. $PbCl_4$

6. Bond energies are quoted in kJ mol⁻¹

   C—C 348   Si—Si 176
   C—Cl 340   Si—Cl 400
   C—O 360   Si—O 419
   C—H 412   Si—H 318

   Which statements are consistent with these values?

a. The bond energies represent the heat liberated when the bond is split.
b. CH_4 is thermally more stable than SiH_4.
c. C—C bonds are the least readily split of those listed.
d. Si should form long Si—Si chains more readily than C forms long C—C chains.
e. Si—O is less readily split by heat than is C—O.

7. Which hydrides do not react with water and are not spontaneously flammable in air?

a. NaH
b. BH_3
c. CH_4
d. SiH_4
e. NH_3

8. Indicate the elements whose atoms do not form positive ions in aqueous solution.

a. B
b. In
c. C
d. Si
e. N

9. Magnesium ribbon was ignited in a porcelain crucible, leaving a grey stain on the bottom. When cold, diluted hydrochloric acid was poured in giving a gas which ignited spontaneously with a sharp detonation.
What is the stain most likely to be?

a. Mg
b. MgO
c. Mg_2O
d. Mg_2Si
e. Si

10. Pick out those substances which have similar covalent structures and physical properties.

a. CaC_2
b. BN
c. Mg_3N_2
d. Diamond
e. SiC

11. Which statements are correct concerning carbon monoxide?

   a. It can be made by oxidizing formic acid with conc. sulphuric acid.
   b. It is a widely used industrial reducing agent.
   c. It combines more readily with haemoglobin than does oxygen.
   d. It is used in the purification of nickel.
   e. It forms carbonyls with several metals.

12. The following statements appear in a textbook written in 1880. Indicate those which you think are still applicable today.

   a. Carbon shows distinctly metallic properties e.g. graphite is a good electrical conductor.
   b. Siloxanes are useless chemical oddities.
   c. There is no commercial use for the element *eka*-silicon (germanium).
   d. Conc. nitric acid spoils more tin than it dissolves.
   e. Spring water containing dissolved chlorides and nitrates should not be drunk unnecessarily (even after boiling) if it is delivered to the house through lead pipes.

13. Which statements concerning tin are correct?

   a. It is extracted by reducing tin dioxide with carbon.
   b. It is attacked by citric acid.
   c. Tin cans can be made by electroplating tin on to sheet steel.
   d. It is possible to reclaim tin from used cans using chlorine.
   e. When a tin can is scratched the iron corrodes faster than the tin.

14. A chloride in acid solution
   i. turns aqueous potassium permanganate colourless;
   ii. gives a white or grey precipitate with aqueous mercury(II) chloride.
   It could be:

   a. CCl_4
   b. $SnCl_2$
   c. $SnCl_4$
   d. $PbCl_2$
   e. $PbCl_4$

15. Which statements concerning silica and silicates are correct?

   a. Some silicates can act as molecular sieves.
   b. Some silicates can act as ion exchange materials.
   c. Soda glass melts at 1269°C.

16. An oxide of lead is warmed with excess nitric acid, leaving a brown residue. The filtrate gives a yellow precipitate when potassium chromate is added.
The oxide could be:

   a. PbO
   b. PbO_2
   c. Pb_3O_4
   d. $2PbO, PbO_2$
   e. $Pb_2[PbO_4]$

17. Indicate the important industrial uses of lead.

   a. Production of calcium plumbate.
   b. Production of antiknock additive for petrol.
   c. Paints.
   d. Drugs.
   e. Primary electrical cells.

18. An element reacts when heated in dry chlorine. The product when added to water fumes and gives an acidic solution. Addition of sodium hydroxide solution to this gives a precipitate which dissolves in excess alkali.
The element could be:

   a. Fr
   b. Mg
   c. Al
   d. C
   e. Sn

19. A solution contains two metal nitrates dissolved in excess nitric acid. Addition of cold dilute hydrochloric acid gives a white precipitate which is removed by filtration. The filtrate gives with sodium hydroxide solution a different white precipitate. The original solution could contain:

   a. $Pb^{2+}_{(aq)}$ and $Ag^+_{(aq)}$
   b. $Pb^{2+}_{(aq)}$ and $Sn^{2+}_{(aq)}$
   c. $Ag^+_{(aq)}$ and $Sn^{2+}_{(aq)}$
   d. $Hg_2^{2+}_{(aq)}$ and $Ag^+_{(aq)}$
   e. $Mg^{2+}_{(aq)}$ and $Sn^{2+}_{(aq)}$

20. From your knowledge of carbon silicon tin and lead, indicate the properties you would expect germanium to have.

   a. Germanium monoxide is amphoteric.
   b. Germanous chloride is an oxidizing agent.
   c. Germanium tetrachloride is readily hydrolysed by water.
   d. There is a germanous sulphide.
   e. There is an ion $Ge^{3+}_{(aq)}$.

GROUP 5 M

1. Ionisation energies for the change $X_{(g)} \to X^{3+}_{(g)}$ are given for group 5M (N to Bi) in **a** to **e**. What value would you expect for $Bi_{(g)} \to Bi^{3+}_{(g)}$

   ΔH(kJ mol^{-1})
   a. +4750
   b. +4920
   c. +5720
   d. +5890
   e. +8830

2. Pauling electronegativity values are given below:

   As 2·0 N 3·0
   C 2·5 O 3·5 Si 1·8
   Ge 1·8 S 2·5

   Deduce the value for phosphorus.

   a. 3·2
   b. 2·6
   c. 2·1
   d. 1·9
   e. 1·7

3. An element forms basic salts listed in a reputable text book as
   $X(OH)_2NO_3$
   $(XO)_2CO_3$
   $(XO)_2SO_4$
   Work out the oxidation state of X in each compound. X could be:

   a. Ra
   b. B
   c. Ge
   d. P
   e. Bi

4. *Property*

Property	N	P	As	Sb	Bi	
Density (g cm^{-3})	0·88 (at $-210°C$)	1·8	(a)	6·6	9·8	a. 5·7
Atomic volume (cm^3 mol^{-1})	15·9	16·9	(b)	18·5	21·3	b. 13·1
Ionisation energy for $X_{(g)} \to X^{3+}_{(g)}$ (kJ mol^{-1})	8850	5880	(c)	4863	4844	5646
Ionic radius of $X^{3+}_{(s)}$ (pm)	—	—	(d)	90	120	69
Standard redox potential for $X_{(s)} + 3H^+_{(aq)} + 3e^- \rightleftharpoons XH_3(g)$ (volts)	+0·27	−0·03	(e)	−0·51	−0·8	e. −0·54

   Which values would you expect to be correct?

5. An element forms a trioxide which dissolves in and reacts with both
   i. sodium hydroxide solution, and
   ii. hydrochloric acid (conc.).
   The element could be:

   a. B
   b. Al
   c. P
   d. As
   e. Sb

6. An element on heating in air forms a trioxide. This compound dissolves in dilute nitric acid, from which deliquescent crystals $X(NO_3)_3, nH_2O$ can be isolated. The element could be

   a. B
   b. In
   c. P
   d. As
   e. Bi

7. Which of the following statements would you expect to be correct concerning the tri-halides of As, Sb and Bi?

   a. The fluorides (XF_3) do not exist.
   b. The melting point of $AsCl_3$ is higher than that of $SbCl_3$.
   c. The boiling point of $SbBr_3$ is lower than that of $SbCl_3$.
   d. $BiCl_3$ has a greater specific conductance for the pure compound than has $AsCl_3$.
   e. $SbBr_3$ will have a planar molecule.

8. A gas, when passed through a glass tube heated at one point by a burner, gives an easily visible silvery sheen on the tube some distance along from the heated part. The gas could be:

   a. NH_3
   b. PH_3
   c. AsH_3
   d. SbH_3
   e. BiH_3

9. A compound when added to water produces a white precipitate which dissolves when excess conc. hydrochloric acid is added.
   The compound could be:

   a. NCl_3
   b. PCl_3
   c. $AsCl_3$
   d. $SbCl_3$
   e. $BiCl_3$

10. Indicate the correct statements concerning the Haber process for the production of ammonia from nitrogen and hydrogen.

    a. It is an equilibrium reaction.
    b. The reaction producing ammonia is exothermic.
    c. A catalyst is essential.
    d. The reaction is normally carried out at 1000 atmospheres and 200°C, these being the optimum conditions.
    e. The process was first used industrially in Germany in 1919.

11. Indicate the correct statements concerning the industrial uses of ammonia.

    a. Some is converted to fertilizers.
    b. Some is converted to ammonium nitrate and used in explosives.

c. Some is converted to nitric acid.
d. Some is converted to urea for use in plastics.
e. Some is catalytically oxidized.

12. Liquid ammonia ionises slightly:
$$2NH_3 \rightleftharpoons NH_4^+ + NH_2^-$$
(cf. $2H_2O \rightleftharpoons H_3O^+ + OH^-$)
Which statements are true concerning reactions in liquid ammonia?

a. Sodamide ($NaNH_2$) acts as an acid.
b. Ammonium chloride acts as a base.
c. Potassamide neutralises ammonium sulphate.
d. $Zn(NH_2)_2$ reacts with KNH_2 to give $K_2[Zn(NH_2)_4]$.
e. Calcium reacts with ammonium ions giving ammonia, calcium ions, and hydrogen.

13. Which statements are correct concerning the oxide NO?

a. The molecule contains an odd number of electrons.
b. The reaction producing it from the elements is exothermic.
c. It is formed during the industrial manufacture of nitric acid from ammonia.
d. Nitrosyl sulphuric acid ($NOHSO_4$) is formed in the chamber process for the manufacture of sulphuric acid.
e. It forms the colourless ion $[FeNO]^{2+}$.

14. Which statements are correct concerning nitrogen dioxide?

a. It disproportionates in alkaline solution to an equimolar mixture of NO_2^- and NO_3^-.
b. The solution in water decomposes liberating brown fumes in air.
c. It dimerises above 200°C.
d. 1 mole of zinc reacted with excess 50 per cent nitric acid gives exactly 2 moles of nitrogen dioxide.
e. Br_2 oxidizes NO_2^- to NO_3^-, whereas NO_2^- oxidizes $2I^-$ to I_2 (all in acid solution).

15. The preparation of phosphine can be considered to proceed according to the equation
$$4P(s) + 3NaOH(aq) + 3H_2O(l) \rightarrow 3NaH_2PO_2(aq) + PH_3(g)$$

a. 1 lost, 1 gained.
b. 2 lost, 2 gained.
c. 3 lost, 3 gained.
d. 1 lost, 3 gained.
e. 3 lost, 1 gained.

This is a redox reaction. Indicate the number of electrons lost from or gained by a phosphorus atom when it is oxidized or reduced.

16. Indicate the properties which are those of pure phosphine PH_3.

   a. It is strongly alkaline in solution.
   b. It forms a salt $PH_4^+ I^-$.
   c. It is spontaneously flammable in air.
   d. It reduces silver nitrate solution.
   e. It decolourises acidified potassium permanganate solution.

17. In which reactions is ammonia (or the ammonium ion) a reducing agent?

   a. Passing dry ammonia over heated copper (II) oxide.
   b. Warming aqueous solutions of ammonium chloride and sodium nitrite.
   c. Mixing excess ammonia gas with chlorine gas.
   d. Thermally decomposing solid ammonium dichromate.
   e. Mixing ammonia and hydrogen bromide gases.

18. These acids are listed in a reputable textbook. Assuming that they exist in aqueous solution, deduce which reactions are likely to occur by working out the oxidation state of the relevant element.

   a. $H_2N_2O_2$ turns $K_2Cr_2O_7(aq)$ green.
   b. $H_4P_2O_5$ decolourises acid $KMnO_4(aq)$.
   c. $H_4As_2O_7$ decolourises bromine water.
   d. $HSbCl_6$ decolourises a solution of iodine in KI.
   e. HBi_2Cl_7 decolourises acid $KMnO_4(aq)$.

19. Deduce which statements are correct.

   a. In the spontaneous decomposition of NO_2^- in acid solution to NO and NO_3^-, no redox reaction occurs.
   b. Hydroxylamine (NH_2OH) can be made by the electrolytic reduction of nitric acid.
   c. In aqueous solution 1 mole of $H_2PO_2^-$ is needed to reduce 2 moles of mercury(II) chloride to mercury.
   d. The conversion of meta-phosphoric acid (HPO_3) to the pyro acid ($H_4P_2O_7$) involves oxidation.
   e. $Ca_3(AsS_3)_2$ is a crystalline compound named calcium thioarsenate.

20. In some versions of the periodic table the element protoactinium (Pa) is placed in group 5M.
Indicate the properties it would need to have to justify this place.

a. It should form Pa^{3+} in aqueous solution.
b. It should form a thermally stable hydride PaH_3.
c. $PaCl_3$ should be a liquid hydrolysed by water.
d. Pa_2O_3 should be soluble in nitric acid giving on crystallisation $Pa(NO_3)_3, nH_2O$.
e. Pa should readily form with oxygen an anion in which the oxidation state of Pa is $+5$.

GROUP 6 M

1. Ionisation energies are given below for the change
$$X_{(g)} \rightarrow X^{2+}_{(g)} + 2e^-$$
Oxygen: $\Delta H = +4700$ kJ mol^{-1}
Sulphur: $\Delta H = +3260$ kJ mol^{-1}
Calcium: $\Delta H = +1740$ kJ mol^{-1}
Deduce the value of ΔH for selenium.

   a. $+1410$ kJ mol^{-1}
   b. $+3020$ kJ mol^{-1}
   c. $+3240$ kJ mol^{-1}
   d. $+4240$ kJ mol^{-1}
   e. $+6260$ kJ mol^{-1}

2. The electron affinity for sulphur is as below:
$$S_{(g)} + 2e^- \rightarrow S^{2-}_{(g)}: \Delta H = +332 \text{ kJ mol}^{-1}$$
Deduce the value of ΔH for
$$O_{(g)} + 2e^- \rightarrow O^{2-}_{(g)}$$

   a. -338 kJ mol^{-1}
   b. -318 kJ mol^{-1}
   c. -163 kJ mol^{-1}
   d. $+326$ kJ mol^{-1}
   e. $+657$ kJ mol^{-1}

3. Pauling electronegativity values are given:
   O 3·5   Te 2·0   F 4·0   N 3·0   Br 2·8
   As 2·0   S 2·5   Cl 3·0   P 2·1
   Deduce the value for Se.

   a. 2·0
   b. 2·4
   c. 2·6
   d. 3·1
   e. 3·8

4. This information has been copied from a reputable textbook but in one or more cases the *order* of values has been deliberately reversed.
In which sets of properties has this been done?

	O	S	Se	Te
a. Number of stable isotopes	3	4	6	8
b. Melting point (°C)	-219	119	217	450
c. Heat of fusion kJ mol^{-1}	0·22	1·42	5·23	17·9
d. Ionisation energy for $X_{(g)} \rightarrow X^+_{(g)}$ kJ mol^{-1}	870	941	1000	1310
e. Atomic volume cm^3 mol^{-1}	20	17	16	13

5. Which statements are correct concerning the different physical forms of sulphur?

   a. The phenomenon can be called both monotropy and polymorphism.
   b. Plastic sulphur consists of intertwined rings and chains of sulphur atoms.
   c. Both crystalline forms consist of S_6 rings packed together.

d. Sulphur crystallising from cold carbon disulphide solution will be monoclinic.
e. Two crystalline forms are in equilibrium at 96°C.

6. 100 cm³ of ozonized air was allowed to stand until all the ozone had decomposed. The final volume was 103 cm³. (Temperature and pressure were constant throughout at 288K and $10^5 Nm^{-2}$.)
Calculate the percentage by volume of ozone in the sample (to the nearest 1 per cent).

a. 2 per cent
b. 3 per cent
c. 6 per cent
d. 9 per cent
e. 12 per cent

7. Which reaction gives the largest volume of oxygen (measured at the same temperature and pressure)?

a. Heating 2 moles of lead dioxide.
b. Heating 2 moles of barium peroxide.
c. Adding 2 moles of sodium peroxide to excess warm water.
d. Electrolysing 2 moles of water.
e. Adding 2 moles of potassium permanganate to excess acidified hydrogen peroxide.

8. Plot these boiling points against the period number of the element forming the hydride for each series. Indicate the compounds in which there are unexpectedly strong intermolecular forces (i.e. hydrogen bonds).

a. CH_4
b. NH_3
c. H_2O
d. H_2Te
e. HCl

Compound	Boiling point °C	Compound	Boiling point °C
H_2O	100	NH_3	−33
H_2S	−62	PH_3	−88
H_2Se	−42	AsH_3	−55
H_2Te	−4	SbH_3	−18
HF	19	CH_4	−161
HCl	−84	SiH_4	−112
HBr	−67	GeH_4	−90
HI	−35	SnH_4	−52

9. A hydride when dissolved in water gives an acidic solution. It could be the hydride of:
   a. Na
   b. Si
   c. P
   d. S
   e. Cl

10. The spontaneous decomposition of aqueous hydrogen peroxide is a redox reaction involving disproportionation. How many electrons are transferred from one molecule of hydrogen peroxide to another in this reaction?
    a. 1
    b. 2
    c. 3
    d. 4
    e. 5

11. Indicate those reactions in which hydrogen peroxide takes part as an oxidizing agent.
    a. Filter paper which has been dipped in lead acetate solution is exposed to hydrogen sulphide and then placed in hydrogen peroxide solution.
    b. With acidified potassium permanganate solution.
    c. With silver nitrate solution to which sodium hydroxide solution has been added.
    d. With acidified iron(II) sulphate solution.
    e. With acidified potassium iodide solution.

12. Hydrogen sulphide is bubbled into separate samples of aqueous solutions. In which cases is the H_2S a reducing agent?
    a. SO_2
    b. $KMnO_4$
    c. $(CH_3COO)_2Pb$
    d. HNO_3(conc)
    e. $Fe_2(SO_4)_3$

13. Information concerning the conversion of sulphur dioxide to sulphur trioxide is given. Which of these consequences follow from the corresponding facts?

    Fact

    a. The reaction is reversible when the gases are confined in a closed vessel.

    b. The reaction producing sulphur trioxide is exothermic.

    c. The reaction between sulphur dioxide and oxygen is slow at room temperature.

    Consequence

    a. 100 per cent conversion of sulphur dioxide to trioxide can never be achieved under these conditions.

    b. A high temperature causes $SO_3(g)$ to decompose to $SO_2(g)$ and $O_2(g)$.

    c. A temperature of about 500°C is used industrially.

d. A catalyst is used.

e. A slight excess of oxygen is used with the sulphur dioxide in the industrial process.

14. Which statements concerning the industrial production of sulphuric acid are correct?

d. This alters the position of equilibrium in favour of sulphur trioxide production.

e. This alters the position of equilibrium in favour of sulphur trioxide production.

a. None is made by the chamber process in the United Kingdom.
b. Some is produced from sulphur extracted from crude oil.
c. The process involving anhydrite gives cement as a by-product.
d. The chamber process only works with very pure sulphur dioxide.
e. The contact process is an example of homogeneous catalysis.

15. A salt in solution
   i. decolourises iodine dissolved in potassium iodide solution;
   ii. produces sulphur dioxide with excess dilute acid.
   It could be:

a. Sodium sulphite.
b. Sodium thiosulphate.
c. Sodium metabisulphite.
d. Sodium hydrogen sulphate.
e. Sodium sulphate.

16. Compounds having these formulae are listed in a reputable textbook. Which of them would you expect to decolourise acidified potassium permanganate solution?

a. $Na_2S_2O_3$
b. Na_2SO_4
c. $Na_2S_2O_4$
d. $Na_2S_2O_6$
e. $K_2S_5O_6$

17. Which of these properties would you expect H_2Se to have?

a. H_2Se burns in excess oxygen giving SeO_2.
b. H_2Se has an unpleasant smell.
c. H_2Se can be made by adding iron(II) selenide to dilute hydrochloric acid.
d. H_2Se is a weak acid in water.
e. H_2Se when passed into cadmium nitrate solution gives an orange-red precipitate.

18. Which of these properties would you expect of selenium?

a. Se exists in at least two crystalline forms, one of which consists of Se_8 rings.
b. Se forms the alum $KAl(SeO_4)_2, 12H_2O$.

c. Se forms a sol in water.
d. Se burns in oxygen giving a basic oxide.
e. Potassium selenite decolourises acidified potassium permanganate solution.

19. Which properties would you expect of these tellurium compounds?

a. H_2Te does not exist.
b. There is a compound Al_2Te_3.
c. TeO_2 is formed when Te burns in air.
d. Sodium tellurite is oxidized in air to sodium tellurate.
e. TeO_2 readily dissolves in water giving a weak acid H_2TeO_3.

20. Which properties would you expect of polonium?

a. Polonium hydride readily decomposes on heating.
b. Na_2PoO_4 exists.
c. Polonium forms a chloride which is readily hydrolysed by water.
d. Polonium forms some salts.
e. Polonium oxide is amphoteric.

GROUP 7 M THE HALOGENS

1. Ionisation energies for the change $X_{(g)} \rightarrow X^+_{(g)}$ are given in kJ mol^{-1}.

	ΔH
F → F$^+$	+1680
Ar → Ar$^+$	+1520
Kr → Kr$^+$	+1350
O → O$^+$	+1310
Cl → Cl$^+$	+1260
Br → Br$^+$	+1140
Na → Na$^+$	+ 494

   Deduce the element which you think will be the least likely to form $X^+_{(g)}$. (Put these values in the periodic table and make reasonable assumptions about those not given.)

   a. Rb
   b. S
   c. F
   d. Cl
   e. Ne

2. Fluorine can be extracted industrially by which methods?

   a. Electrolysis of fused HF + KF mixtures.
   b. Reduction of F$^-$ to F.
   c. Chemical oxidation of HF.
   d. Electrolysis of conc. KF$_{(aq)}$.
   e. Thermal decomposition of HF.

3. Which compound would you expect to have the largest permanent dipole moment?

   a. CH$_3$F
   b. CH$_3$Cl
   c. CH$_3$Br
   d. CH$_3$I
   e. CF$_4$

4. Given that fluorine has a smaller van der Waals radius than iodine, indicate the statements which you would expect to be correct.

   a. Fluorine is less electronegative than iodine.
   b. The ionisation energy of $F_{(g)} \rightarrow F^+_{(g)}$ is less than that of $I_{(g)} \rightarrow I^+_{(g)}$.
   c. The electron affinity of fluorine is greater than that of iodine.
   d. Fluorine is more reactive towards metals and hydrogen than is iodine.
   e. The hydrogen—halogen bond in HF is 60 per cent ionic in character, and that in HI is 5 per cent.

5. Which statements concerning the production of chlorine are correct?

   a. 2 moles of potassium permanganate with excess conc. hydrochloric acid liberate 6 moles of chlorine.
   b. A solution containing 1 mole each of ClO^- and Cl^- liberates 1 mole of chlorine with excess hydrochloric acid.
   c. 2 moles of hydrogen chloride passed with air over a hot catalyst give 1 mole of chlorine.
   d. 2 moles of sodium chloride and 2 moles of calcium chloride when electrolysed in a Downs cell liberate 4 moles of chlorine.
   e. 2 moles of sodium chloride electrolysed in a mercury cell liberate 2 moles of sodium hydroxide and 2 moles of chlorine.

6. What is the minimum number of moles of sodium hydroxide which can be used to absorb 1 mole of chlorine gas?

   a. 1
   b. 2
   c. 3
   d. 4
   e. 5

7. A white solid on warming gently with conc. sulphuric acid gives a mixture of white and coloured fumes.
   The solid could be:

   a. CaF
   b. LiCl
   c. NaBr
   d. KI
   e. KHF_2

8. The dipole moment of hydrogen chloride gas is 1·03 Debye units.
   What value would you expect for hydrogen bromide?

   a. 1·91
   b. 1·47
   c. 1·03
   d. 0·79
   e. 0·00

9. A gaseous compound dissolves in water, thus becoming virtually fully ionised. It could be:

   a. HCl
   b. HBr
   c. HI
   d. H_2S
   e. H_2Se

Use the values given below to answer Questions 10 to 13, assuming that all ions involved are in molar solution (including H+ where relevant). Standard redox potentials at 25°C are given in volts.

$$F_2(g) + 2e^- \rightleftharpoons 2F^-_{(aq)} \qquad E^\circ \text{ (volts)} +2\cdot87$$
$$MnO_4^-{(aq)} + 8H^+_{(aq)} + 5e^- \rightleftharpoons Mn^{2+}_{(aq)} + 4H_2O(l) \qquad +1\cdot51$$
$$Cl_2(aq) + 2e^- \rightleftharpoons 2Cl^-_{(aq)} \qquad +1\cdot40$$
$$Cr_2O_7^{2-}{(aq)} + 14H^+_{(aq)} + 6e^- \rightleftharpoons 2Cr^{3+}_{(aq)} + 7H_2O(l) \qquad +1\cdot33$$
$$O_2(g) + 4H^+_{(aq)} + 4e^- \rightleftharpoons 2H_2O(l) \qquad +1\cdot23$$
$$2IO_3^-{(aq)} + 12H^+_{(aq)} + 10e^- \rightleftharpoons I_2(s) + 6H_2O(l) \qquad +1\cdot19$$
$$Br_2(aq) + 2e^- \rightleftharpoons 2Br^-_{(aq)} \qquad +1\cdot09$$
$$I_2(aq) + 2e^- \rightleftharpoons 2I^-_{(aq)} \qquad +0\cdot62$$
$$Na^+_{(aq)} + e^- \rightleftharpoons Na(s) \qquad -2\cdot71$$

10. Which oxidizing agents would you expect to liberate chlorine from a solution of chloride ions?
   a. O_2
   b. $K_2Cr_2O_7$ in acid
   c. $KMnO_4$ in acid
   d. F^-
   e. Br^-

11. Which would you NOT expect chlorine to oxidize?
   a. I^-
   b. Cr^{3+}
   c. Mn^{2+}
   d. F^-
   e. Br^-

12. Which would you expect to liberate iodine from an acidified solution of potassium iodate?
   a. MnO_4^- in acid
   b. $Cr_2O_7^{2-}$ in acid
   c. I^-
   d. Br^-
   e. O_2

13. Which would you expect to liberate bromine from an acidified solution of bromide ions?
   a. O_2
   b. Cl_2
   c. $MnSO_4(aq)$
   d. I_2
   e. Na

14. Bromine is produced industrially in this country by which methods?
   a. Electrolysis of sea water.
   b. Passing chlorine into acidified sea water.
   c. Bubbling oxygen into acidified sea water.
   d. Using the reaction $2Br^-(aq) \rightarrow Br_2(g) + 2e^-$
   e. Reducing the bromide ion.

15. Which statements concerning iodine are correct?
    a. The extraction of iodine from caliche involves the reduction of $IO_3^-(aq)$.
    b. Iodine oxidizes sodium thiosulphate.
    c. Iodine forms the ion $I_3^-(aq)$.
    d. $HI_{(aq)}$ reduces $NO_3^-{}_{(aq)}$ when the latter is in concentrated acidic solution.
    e. Iodine forms a basic oxide I_2O_5.

16. A gas when passed through conc. nitric acid liberates nitrogen dioxide. The gas could be:
    a. HF
    b. HCl
    c. HBr
    d. HI
    e. H_2S

17. A white solid is heated with conc. sulphuric acid, liberating white fumes. These turn drops of water on the inside of the test tube milky.
    The solid could be:
    a. NaF
    b. CaF_2
    c. NaCl
    d. $NaClO_3$
    e. NaBr

18. In a mixture prepared for analysis, damp potassium iodide was mixed with another compound. On long standing, iodine was liberated.
    The other compound could have been:
    a. Copper(II) sulphate.
    b. Aluminium potassium sulphate.
    c. Sodium chloride.
    d. Zinc sulphate.
    e. Sodium hydroxide.

19. These properties of fluorine and its compounds are correctly stated.
    Which agree with those you can predict from your knowledge of chlorine bromine and iodine?
    a. CaF_2 is practically insoluble in water.
    b. The hydride is a weak acid in aqueous solution.
    c. The hydride attacks silica.
    d. Fluorine forms an oxide.
    e. Fluorine oxidizes gold.

20. Indicate the properties you would expect the new element astatine to have.
    (The element was first reported in 1942.)
    a. It forms At^-
    b. It is radioactive.
    c. It is a liquid at room temperature.
    d. It forms a hydride which is covalent in the absence of water.
    e. It could be prepared by electrolysis of NaAt(aq).

TRANSITION METALS

1.   *Element*

	(1)	(2)	(3)	(4)	(5)
$M_{(g)} \to M_{(g)}^+$	6	7	5	22	7
$M_{(g)} \to M_{(g)}^{2+}$	25	21	53	63	21
$M_{(g)} \to M_{(g)}^{3+}$	53	51	125	127	49
$M_{(g)} \to M_{(g)}^{4+}$	173	99	224	225	92
$M_{(g)} \to M_{(g)}^{5+}$	327	164	363	350	191

a. Element 1
b. Element 2
c. Element 3
d. Element 4
e. Element 5

All values are in 10^5 J mol^{-1}.

These are the energies needed to produce various degrees of ionisation in different elements of approximately equal atomic radii.

Deduce which of the elements are probably transition metals.

2. Which are the electron arrangements of transition elements?

a. 2 8 10 2
b. 2 8 13 2
c. 2 8 18 7
d. 2 8 18 21
e. 2 8 18 18 8

3. Which are the electron arrangements of lanthanides?

a. 2 8 18 8 1
b. 2 8 18 10 2
c. 2 8 18 21 8 3
d. 2 8 18 32 18 8
e. 2 8 18 34 14 2

4. According to some authorities Zn, Cd and Hg are not transition metals because of their electron structures.

Indicate the properties you would expect these elements to have if they were to be classed as transition metals.

a. They would appear in an anion in at least one compound.
b. They would not form cations.
c. They would exist in only one oxidation state in their compounds.
d. They would form ammines with ammonia, and similar compounds with other ligands.
e. The elements or their compounds might be catalysts.

5. An element has all of these properties:
   i. It forms at least one coloured compound.
   ii. It exists in more than one oxidation state in its compounds.
   iii. It can form part of an anion.
   The element could be:

   a. Fe
   b. Hg
   c. Sn
   d. Pb
   e. Nothing but a transition metal.

6. Indicate the ions or molecules which you consider to be complexes.

   a. VO_3^-
   b. $[Fe(CN)_6]^{3-}$
   c. CrO_4^{2-}
   d. $Pt(NH_3)_2Cl_4$
   e. $[Co(NH_3)_5SO_4]^-$

7. Indicate the groups which you would expect to act as ligands in complexes.

   a. NH_3
   b. NH_4^+
   c. H_2O
   d. Cl^-
   e. $NH_2CH_2CH_2NH_2$

8. Which of these salts would you expect to give complex ions containing metals when they are dissolved in water?

   a. $MgNH_4PO_4$
   b. $NaNH_4HPO_4$
   c. $Cu(NH_4)_2(SO_4)_2, 6H_2O$
   d. $Fe(NH_4)(SO_4)_2, 12H_2O$
   e. $Cu(NH_3)_2Cl$

9. In which cases are the oxidation states given correctly?

Compound	Oxidation state
a. K_2TiO_3	Ti^{IV}
b. NH_4VO_3	V^V
c. $(NH_4)_2MoO_4$	Mo^{VI}
d. $Ni(CO)_4$	Ni^0
e. $MnCl_6^{2-}$	Mn^{-IV}

10. In which reactions has oxidation occurred?

    a. $V_2O_5 \rightarrow Na_3VO_4$
    b. $K_2MnO_4 \rightarrow KMnO_4$
    c. $[ZnO_2]^{2-} \rightarrow Zn^{2+}$
    d. $Cu^{2+} \rightarrow [Cu(NH_3)_4]^{2+}$
    e. $Fe^{2+} \rightarrow [Fe(CN)_6]^{3-}$

11. To convert insoluble chromium(III) hydroxide to soluble sodium chromate, the precipitate is warmed with aqueous sodium hydroxide containing

a. H_2O_2
b. Na_2O_2
c. Na_2SO_4
d. Na_2SO_3
e. NaH_2PO_2

12. Which reagents are likely to convert $Cr_2O_7^{--}$ to CrO_4^{--} in aqueous solution?

a. Na_2SO_3(aq)
b. H_3PO_4(aq)
c. SO_2(aq)
d. NaOH(aq)
e. HNO_3 (conc)

13. In the presence of carbon dioxide MnO_4^{2-}(aq) is converted to MnO_4^{-}(aq) and MnO_2. How many moles of K_2MnO_4 give two moles of $KMnO_4$ under these conditions?

a. 1
b. 2
c. 3
d. 4
e. 5

14. Indicate the properties you would expect of potassium hexacyanoferrate(II) in aqueous solution.

a. 1 mole of the compound gives 5 moles of ions.
b. Addition of sodium hydroxide solution gives a green precipitate of iron(II) hydroxide.
c. Passage of hydrogen sulphide into an acidified solution gives black iron(II) sulphide.
d. Addition of silver nitrate solution precipitates white silver cyanide.
e. The compound does not appear in the official Poisons List.

15. Using only one mole of oxidizing agent under suitable conditions indicate which will oxidize the largest number of moles of iron(II) sulphate in solution.

a. $KMnO_4 + H_2SO_4$ (dil)
b. $KMnO_4 + NaOH$ (dil)
c. $K_2Cr_2O_7 + H_2SO_4$ (dil)
d. $H_2O_2 + H_2SO_4$ (dil)
e. Cl_2

16. A dilute solution of a compound in water has these properties.
   i. Hydrogen sulphide readily gives a black precipitate.
   ii. Iron filings displace copper.
   iii. The solution is light blue in colour.
The compound could be:

a. $Cu(NH_3)_4SO_4, H_2O$
b. $Cu(NH_4)_2(SO_4)_2, 6H_2O$
c. $Cu(NH_3)_4(NO_3)_2, 2H_2O$
d. $Cu(NH_3)_4(CNS)_2$
e. Tetramminecopper(II) nitrate.

TRANSITION METALS

17. To dissolve silver chloride unaffected by light from a photographic film, a fixer is employed.
    Indicate the reagents which you think would work, assuming that they do not affect the film base.

    a. KCN(aq)
    b. NH_3(aq)
    c. $Na_2S_2O_3$(aq)
    d. Na_2CrO_4(aq)
    e. HCl(dil)

18. The electron structure of copper is listed as 2 8 18 1.
    Copper forms a complex in aqueous solution with ethylene diamine of formula $[Cu(NH_2CH_2CH_2NH_2)_2]^{2+}SO_4^{2-}$
    Indicate which statements you would expect to be correct, knowing that the complex does not give a precipitate with hydrogen sulphide in aqueous solution.

    a. The compound is not a complex as copper does not have a rare gas electron structure.
    b. The complex ion is more readily formed in solution than $[Cu(H_2O)_4]^{2+}$.
    c. Addition of sodium hydroxide solution would be expected to give a precipitate of copper(II) hydroxide.
    d. The oxidation state of copper is Cu^{II}.
    e. The compound is a chelate complex.

19. 0·01 mole of a compound was dissolved in water and the solution passed through a high capacity cation exchange column (which exchanges all the positive ions for $H^+_{(aq)}$).
    The emergent solution required 40·0 ml M NaOH in a titration.
    The compound could have been

    a. $CuSO_4, 5H_2O$
    b. $Cu(NH_3)_4SO_4, H_2O$
    c. $CuSO_4, (NH_4)_2SO_4, 6H_2O$
    d. $Cu(NO_3)_2, 3H_2O$
    e. $CuCl_2, H_2O$

20. A compound has the composition by weight:
    Co 23·5 per cent
    Cl 42·5 per cent
    NH_3 34·0 per cent
    The molar mass is 251.
    1·000 g in solution gives 1·144 g of silver chloride when excess silver nitrate is added.
    The compound could be:

    a. $Co(NH_3)_5Cl_3$
    b. $[Co(NH_3)_5Cl]^{2+}$ $2Cl^-$
    c. $[Co(NH_3)_5]^{3+}$ $3Cl^-$
    d. $[Co(NH_3)_5Cl_2]^+$ Cl^-
    e. Chloropentamminecobalt(III) chloride.

ALKANES, ALKENES AND ALKYNES

1. Solid sodium propanoate is heated with a mixture of calcium hydroxide and sodium hydroxide.
   Which statements are correct?
   a. The product is mainly methane.
   b. The product is mainly ethane.
   c. The product is pure propane.
   d. The method is a good industrial way of making an alkane.
   e. The reaction can be named decarboxylation.

2. Sodium butanoate solution is electrolysed in a suitable apparatus.
   The product at the anode is mainly:
   a. Butane.
   b. Butane + carbon dioxide.
   c. Pentane.
   d. Hexane.
   e. Hexane + carbon dioxide.

3. A primary alcohol has the elements of water removed from it.
   Which statements are correct?
   a. A double bond results.
   b. The process can also be named dehydrogenation.
   c. The reaction can be performed using hot conc. sulphuric acid or hot aluminium oxide.
   d. Under suitable conditions, the reaction can be reversed.
   e. The product is an alkyne.

4. Butan-1-ol is dehydrated in a suitable apparatus.
   The product is:
   a. Butane.
   b. Butylene.
   c. But-1-ene.
   d. But-2-ene.
   e. But-1-yne.

5. Which statements are correct concerning ethylene?
   a. It is made in large quantities by hydrogenating acetylene.
   b. It can be made in the laboratory by dehydrating ethanol.
   c. Refluxing iodoethane with aqueous sodium hydroxide gives ethylene.
   d. It can be polymerised.
   e. It combines readily with bromine.

6. Which statements are correct concerning acetylene (ethyne)?
   a. Much is made industrially by adding water to calcium carbide.

                                                                             b. A lot is polymerised to give benzene.
                                                                             c. It reduces ammoniacal silver nitrate solution to silver.
                                                                             d. It can be combined with hydrogen chloride to give (eventually) PVC.
                                                                             e. It is an endothermic compound.

7. When equal volumes of ethane and chlorine are mixed in ultra-violet light the organic product is:
      a. pure chloroethane.
      b. mainly chloroethane.
      c. pure 1,2-dichloroethane.
      d. mainly 1,2-dichloroethane.
      e. mainly hexachloroethane.

8. The reaction between equal volumes of propylene and hydrogen chloride under suitable conditions
      a. gives mainly 1-chloropropane.
      b. gives mainly 2-chloropropane.
      c. gives mainly 2-chloropropene.
      d. is called substitution.
      e. may be carried out by passing the gases over a hot catalyst.

9. The reaction between equal volumes of acetylene and hydrogen chloride under suitable conditions
      a. can be called addition.
      b. can be called polymerisation.
      c. gives mainly monochloroethylene.
      d. gives mainly vinyl chloride.
      e. gives mainly 1,1-dichloroethane.

10. The reaction between concentrated sulphuric acid and a carbon-carbon double bond
      a. can be called addition.
      b. can be called substitution.
      c. involves oxidation of the organic compound (excluding any charring which occurs).
      d. can be used to make 'Teepol' (a sodium alkyl sulphate detergent).
      e. can be used to make synthetic alcohols.

11. The reaction between hydrogen and a carbon-carbon double bond
      a. can be called addition.
      b. can be called hydrogenation.
      c. is used in the manufacture of margarine.
      d. needs a catalyst.
      e. is usually done at room temperature and pressure.

12. Which reagents attack a carbon-carbon double bond under suitable conditions in the laboratory?

   a. $O_3(g)$
   b. $Cl_2(g)$
   c. $KCN(aq)$
   d. $HClO(aq)$
   e. $NH_3(g)$

13. Which reagents attack a carbon-carbon treble bond under suitable conditions in the laboratory?

   a. $HCl(g)$
   b. $Cl_2(g)$
   c. $KCN(aq)$
   d. $KMnO_4$ in acid solution
   e. $H_2(g)$

14. Which plastics are addition polymers?

   a. Polytetrafluorethylene.
   b. Polythene.
   c. Polystyrene.
   d. Perspex.
   e. Polyvinyl chloride.

15. An organic compound rapidly decolourises bromine water and acid potassium permanganate solution. Its formula could be:

   a. CH_4
   b. C_2H_4
   c. C_2H_6
   d. C_4H_6
   e. C_6H_{14}

16. An organic compound on shaking with bromine water and acid potassium permanganate solution decolourises them both. It gives a precipitate with ammoniacal silver nitrate solution. The compound could be:

   a. C_2H_2
   b. C_2H_4
   c. C_2H_6
   d. C_3H_4
   e. C_4H_6

17. Which of these compounds would you expect to decolourise both bromine water and acid potassium permanganate solution?

   a. Vinyl chloride.
   b. Trichlorethylene.
   c. Benzene.
   d. Calor gas (propane + butane).
   e. Propylene.

18. An organic compound does not decolourise bromine water and has no effect on acid potassium permanganate solution. It could be:

   a. C_2H_2
   b. C_2H_4
   c. C_2H_6
   d. C_3H_4
   e. C_4H_6

19. Which properties would you predict for the following compound:

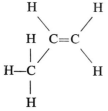

a. It is a liquid at room temperature and pressure.
b. It can be polymerised.
c. It decolourises acid potassium permanganate solution.
d. It decolourises bromine and bromine water.
e. It gives a precipitate with ammoniacal silver nitrate solution.

20. Which statements are correct concerning carbon-carbon bonds?

a. Double bonds are more reactive towards most reagents than single bonds.
b. There is a higher electron density between the atoms in a double bond than in a single bond.
c. The distance between the nucleii of the carbon atoms is shorter in double bonds than in single bonds.
d. The hydrogen and carbon atoms in ethylene are all in the same plane.
e. Compounds of the type

can exist in both *cis* and *trans* forms.

ALCOHOLS AND ALKYL HALIDES

1. Which methods are suitable for the preparation of primary alcohols?

   a. Hydrolysis of a primary alkyl iodide.
   b. Hydrolysis of a secondary alkyl bromide.
   c. Hydrolysis of an ester of a primary alcohol.
   d. Reduction of an acid using sodium and alcohol.
   e. Hydrolysis of an acyl chloride.

2. A liquid liberates hydrogen when sodium is added. On evaporating the resulting mixture, a white solid is left. The liquid could be:

   a. water.
   b. propan-1-ol.
   c. propan-2-ol.
   d. acetic acid.
   e. propanal

3. 4·1 g of 1-iodobutane are refluxed with 12·5 ml of 10 per cent potassium hydroxide solution (i.e. 10 g KOH in 100 ml solution). Which statements are correct?

   a. There is 0·0223 mole of the alkyl halide.
   b. There is 0·0300 mole of alkali.
   c. For practical purposes, neither reactant is in excess.
   d. 0·0223 mole of butan-1-ol is produced.
   e. 1·65 g of the alcohol are produced.

4. A compound is added separately to
   i. $PCl_3(l)$
   ii. $PCl_5(s)$
   iii. $SOCl_2(l)$
   In each case hydrogen chloride is liberated.
   The compound could be:

   a. water.
   b. methanol.
   c. benzene.
   d. acetone.
   e. diethyl ether.

5. Ethylene can be converted industrially to ethanol. An identical reaction to this can be used to prepare propan-2-ol from

   a. propylene.
   b. cyclo-propane.
   c. propyne.
   d. polypropylene.
   e. propane.

6. Calculate the number of moles of ethanol which can be oxidized by 1 mole of potassium dichromate in excess acid if a stream of air is continually blown through the mixture.

   a. 1
   b. 2
   c. 3
   d. 4
   e. 5

7. Calculate the number of moles of ethanol which can be oxidized by refluxing it under a water-cooled condenser with 2·0 moles of potassium permanganate (in excess dilute acid).

   a. 1·1
   b. 1·7
   c. 2·1
   d. 2·5
   e. 3·9

8. The organic products from the reactions in Questions 6 and 7 are different because:

   a. potassium permanganate is the poorer oxidizing agent.
   b. potassium dichromate gives a higher temperature.
   c. acetaldehyde can only be oxidized by potassium permanganate.
   d. the reaction with potassium permanganate is faster.
   e. the aldehyde is removed before it can be further oxidized.

9. The product formed when refluxing propan-2-ol with excess acidified potassium dichromate solution is:

   a. acetone.
   b. acetaldehyde.
   c. acetic acid.
   d. propanal.
   e. There is little or no reaction.

10. The product formed when refluxing 2-methylpropan-2-ol with excess acidified potassium permanganate is:

    a. propanone.
    b. propanal.
    c. propanoic acid.
    d. acetaldehyde.
    e. There is little or no reaction.

11. CH_2OH
    |
    $CHOH$ This compound
    |
    CH_2OH

    a. is a trihydric alcohol.
    b. is glycol.
    c. is immiscible with water.
    d. is a by-product of soap manufacture.
    e. forms an ester with 3 moles of nitric acid per mole of compound.

12. A compound when added to a warm mixture of sodium hypochlorite and potassium iodide solutions gives yellow crystals of iodoform (triodomethane) after several minutes. The compound could be:

    a. ethanol.
    b. acetone.
    c. acetic acid.
    d. propanone.
    e. propan-2-ol.

13. A compound was warmed with sodium hydroxide solution.
The solution formed was acidified with dilute nitric acid and addition of silver nitrate solution gave a distinctly yellow precipitate.
The compound could have been:

   a. chloromethane.
   b. acetyl chloride.
   c. iodoethane.
   d. 2-iodopropane.
   e. potassium iodide.

14. Indicate the reagents which you would expect to react with 1-iodopropane, and the major product of the reaction.

Reagent	Product
(f) Ca(OH)$_2$(aq)	(p) Propane
(g) Alcoholic KOH	(q) Propylene
(h) Alcoholic KCN	(r) 1-hydroxypropane
(i) Dissolving metal reduction	(s) 1-cyanopropane

   a. f giving p
   b. f giving r
   c. g giving q
   d. h giving s
   e. i giving p

15. 1·8 g of red phosphorus, 10·9 g of iodine and 4·0 g of ethanol are refluxed together to produce iodoethane.
Which statements are correct?
[Assume the equation
$$2P + 6C_2H_5OH + 3I_2 \rightarrow 6C_2H_5I + 2H_3PO_3]$$

   a. 0·058 mole of P is used.
   b. 0·0856 mole of I_2 is used.
   c. 0·087 mole of C_2H_5OH is used.
   d. From the equation the ratio of moles of P:I_2:C_2H_5OH is 2:3:6.
   e. There is an excess of iodine and ethanol.

16. 6·0 g of propan-1-ol were refluxed with 13·0 g of solid potassium bromide and 10·8 g of conc. sulphuric acid. 7·7 g of pure 1-bromopropane were produced.
Which statements are correct?
[Assume the equation
$$C_3H_7OH + KBr + H_2SO_4 \rightarrow C_3H_7Br + KHSO_4 + H_2O]$$

   a. 0·100 mole of propanol was used.
   b. 0·109 mole of KBr was used.
   c. 0·056 mole of H_2SO_4 was used.
   d. There was excess KBr and H_2SO_4.
   e. The yield was approximately 63 per cent.

17. Which statements are correct concerning the reactions between ethanol and inorganic acids?

   a. Ethanol can form the ester ethyl hydrogen sulphate.
   b. Excess ethanol warmed with conc. sulphuric acid gives diethyl ether.

ALCOHOLS AND ALKYL HALIDES

c. Ethanol warmed with excess conc. sulphuric acid gives ethylene.
d. Ethanol can be dehydrated using hot syrupy phosphoric acid.
e. Ethanol refluxed with excess conc. hydriodic acid gives ethyl iodide.

18. Indicate the properties you would expect of

H H
| |
H—C—C—H
| |
Br OH

a. Refluxing with KOH(aq) gives glycol.
b. Refluxing with alcoholic KCN give.

c. Refluxing with NaOH(aq) gives ethylene.
d. Reaction with sodium gives hydrogen.
e. Refluxing with glacial acetic acid gives an ester.

19. Methyl iodide dissolved in carefully dried diethyl ether is mixed with excess magnesium turnings under a water-cooled condenser. When the reaction is complete the ether layer is separated. Which reactions of the Grignard organo-metallic compound so produced can be made to occur by adding the necessary reagent to the ether layer?

a. On adding water, methane is evolved.
b. On warming with dilute hydrochloric acid, methanol is produced.
c. On adding solid carbon dioxide followed by hydrolysis with water, acetic acid is formed.
d. On adding acetaldehyde and then hydrolysing, a secondary alcohol is obtained.
e. On adding acetone and then hydrolysing, 2-methylpropan-2-ol is produced.

20. Which statements concerning the modern production of organic chemicals are correct?

a. Dichloromethane is made by chlorinating methane with a catalyst or in ultra-violet light.
b. Methanol is obtained largely by the destructive distillation of wood.
c. C_2H_4O is widely used in industrial syntheses.
d. Ethanol can be converted to acetaldehyde either by oxidation or by dehydrogenation.
e. The ratio of tonnages of chemicals made from oil, coal and coke, and by fermentation is approximately 20:60:1.

ALDEHYDES, KETONES, ACIDS AND ESTERS

The following information may be useful:
Boiling points in °C

Formic acid	101	Formaldehyde	−21	Ethanol	78
Acetic acid	118	Acetaldehyde	20	Butan-1-ol	118
Butanoic acid	163	Butyraldehyde (or Butanal)	75	Propan-1-ol	97
Propyl acetate	101	Methyl ethyl ketone (or Butanone)	80	Acetyl chloride	52
				Propanoyl chloride	80

Thionyl chloride	78	Sulphuryl chloride	70
Phosphorus trichloride	76	Phosphorus pentachloride sublimes at	163
Phosphorus oxychloride	107		

Constant boiling mixtures
70% Propan-1-ol/30% water boils at 88°C.
31% Ethanol/69% ethyl acetate boils at 72°C.

1. 1 mole of butan-2-ol is refluxed with potassium dichromate solution in excess sulphuric acid. The minimum number of moles of potassium dichromate needed is:
   a. 0·17
   b. 0·24
   c. 0·33
   d. 0·47
   e. 1·00

2. The product in Question 1 is most likely to be:
   a. Propanone.
   b. Butanone.
   c. Propanal.
   d. Propanoic acid.
   e. Butanal.

3. Ethanol is refluxed under a water-cooled condenser with excess acidified potassium dichromate solution for an hour.
   What would you expect the final product to be?
   a. Formaldehyde
   b. Formic acid
   c. Acetaldehyde
   d. Acetic acid
   e. Carbon dioxide

4. An organic compound, when added to boiling Fehling's solution [complexed copper(II) hydroxide] gives a red precipitate. When added to Tollens' reagent [ammoniacal silver(I) nitrate solution] it gives a black precipitate.
   The compound could be:
   a. C_2H_5OH
   b. CH_3CHO
   c. CH_3COCH_3
   d. CH_3COOH
   e. Glucose

5. A compound is oxidized with hot acidified potassium permanganate solution and the product drawn in a stream of air through
   i. 2,4-dinitrophenylhydrazine, giving an orange precipitate.
   ii. Tollens' reagent, giving no precipitate.
   iii. hot Fehling's solution, giving no precipitate.
   The compound could be:

   a. Methanol
   b. Ethanol
   c. Propan-2-ol
   d. Butan-2-ol
   e. Pentan-3-ol

6. 1,1-Dichloropropane is refluxed under a water-cooled condenser for some time with sodium hydroxide solution. The product is:

   a. Propanal
   b. Propanol
   c. Propanoic acid
   d. 1-Chloro-1-hydroxypropane
   e. 1,1-Dihydroxypropane

7. Which reagents add onto a
   $\diagdown$
   $\quad$C=O group in most
   $\diagup$
   aldehydes and ketones?

   a. NH_3
   b. HCN
   c. HCl
   d. $NaHSO_4$
   e. H_2

8. An organic compound gives white crystals with an alkaline solution of hydroxyammonium chloride, an orange precipitate with hot Fehling's solution and a black precipitate with Tollens' reagent. On reduction with sodium and ethanol the original compound could give:

   a. Butan-1-ol
   b. Butanal
   c. Butan-2-ol
   d. Butanoic acid
   e. Butane

9. Which compound has all of these properties?
   i. It does not condense with 2,4-dinitrophenylhydrazine.
   ii. It does not condense with hydroxyammonia.
   iii. It ionises in water sufficiently to visibly affect an indicator.

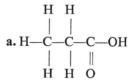

b.
$$\text{H-}\underset{\underset{\text{OH}}{|}}{\overset{\overset{\text{H}}{|}}{\text{C}}}-\underset{\underset{\text{OH}}{|}}{\overset{\overset{\text{H}}{|}}{\text{C}}}\text{-H}$$

c.
$$\text{H-}\underset{\underset{\text{H}}{|}}{\overset{\overset{\text{H}}{|}}{\text{C}}}-\underset{\underset{\text{O}}{\|}}{\text{C}}-\underset{\underset{\text{H}}{|}}{\overset{\overset{\text{H}}{|}}{\text{C}}}-\underset{\underset{\text{H}}{|}}{\overset{\overset{\text{H}}{|}}{\text{C}}}\text{-OH}$$

d.
$$\text{H-}\underset{\underset{\text{H}}{|}}{\overset{\overset{\text{H}}{|}}{\text{C}}}-\underset{\underset{\text{O}}{\|}}{\text{C}}-\underset{\underset{\text{H}}{|}}{\overset{\overset{\text{H}}{|}}{\text{C}}}\text{-H}$$

e.
$$\underset{O\diagup\diagdown OH}{\overset{O\diagdown\diagup OH}{\underset{C}{\overset{C}{|}}}}$$

10. Which statements concerning the preparation and uses of sa.ts of acetic acid are correct?

   a. The best way to make sodium acetate is to add sodium to glacial acetic acid.
   b. Solid sodium acetate when heated with soda lime gives butane.
   c. Ammonium acetate (used to make acetamide) is best made by refluxing acetic acid and conc. ammonia solution.
   d. Acetyl chloride and sodium acetate when heated together give an acid anhydride.
   e. Calcium acetate when heated strongly gives acetone in good yield.

ALDEHYDES, KETONES, ACIDS AND ESTERS

11. An organic compound
  i. effervesces with sodium hydrogen carbonate solution.
  ii. decolourises bromine water rapidly.
  iii. decolourises acid potassium permanganate solution rapidly.
  It could be:

a. 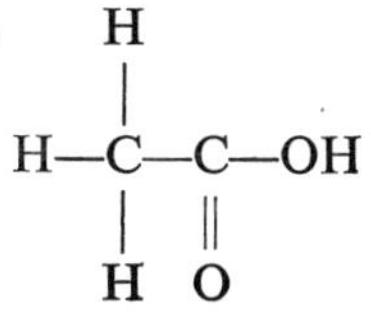
$$\begin{array}{c} H \\ | \\ H-C-C-OH \\ | \;\; \| \\ H \;\; O \end{array}$$

b.
$$\begin{array}{c} \\ H C-OH \\ \diagdown \diagup \| \\ C=C O \\ \diagup \diagdown \\ H H \end{array}$$

(C=C with H, H on left and COOH, H on right)

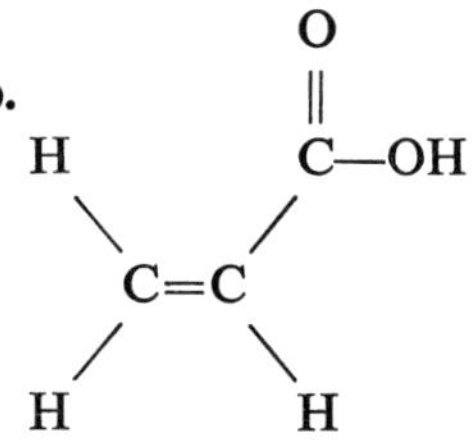

c.
(C=C with phenyl, H on left and COOH, H on right)

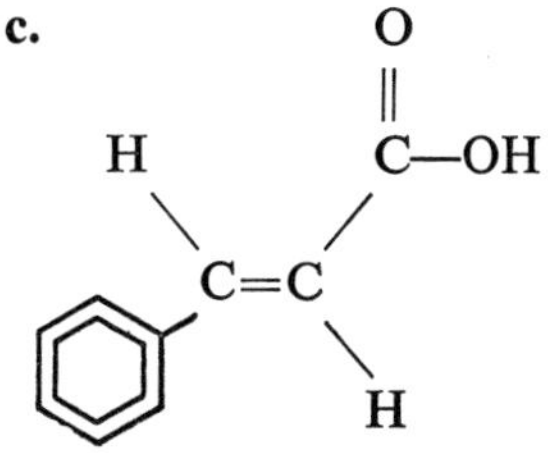

d. 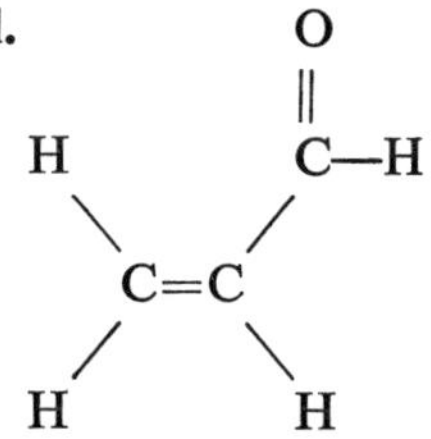
(C=C with H, H on left and CHO, H on right)

e. Oleic acid

12. An organic acid has a molar mass between 100 and 130. 0·13 g of it requires 25·0 ml of 0·1 M NaOH in a titration. The acid could be:

a. 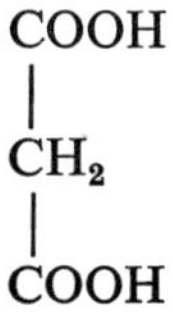 COOH
   |
   CH$_2$
   |
   COOH

b. COOH
   |
   (CH$_2$)$_2$
   |
   COOH

c. COOH
   |
   C=O
   |
   CH$_2$OH

d. CH$_3$CH$_2$CH$_2$CH$_2$COOH

e. COOH
   |
   (CH$_2$)$_4$
   |
   COOH

13. It is intended to convert glacial acetic acid into the corresponding acyl chloride by refluxing with a suitable reagent and distilling off the acyl chloride.
    Which reagents would you expect to be suitable?

    a. SOCl$_2$
    b. PCl$_3$
    c. POCl$_3$
    d. PCl$_5$
    e. SO$_2$Cl$_2$

14. An organic liquid gives an acidic solution when added to water. When added to ethanol with shaking, the liquid gives a sweet smelling oil.
    The liquid could be:

    a. CH$_3$COCl

    b. CH$_3$CH$_2$COCl

    c.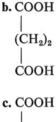

    d. CH$_3$CH$_2$CO
        \
         O
        /
       CH$_3$CH$_2$CO

    e. C$_{17}$H$_{35}$COOH

15. 3·6 g of the ester of propan-1-ol and acetic acid are hydrolysed with 1·69 g of solid sodium hydroxide dissolved in water. Which statements are correct?

a. The ester is $CH_3CH_2CH_2COOCH_3$.
b. There is 0·0353 mole of ester.
c. There is 0·0422 mole of sodium hydroxide present initially.
d. 0·0706 mole of alcohol is produced.
e. 0·276 g of sodium hydroxide is unused.

16. At the end of the reaction in Question 15, if excess sulphuric acid is added and the mixture distilled using a fractionating column, the first distillate will be:

a. pure propan-1-ol.
b. water.
c. acetic acid.
d. sulphuric acid.
e. an azeotropic mixture of propan-1-ol and water.

17. Which of these esters would you expect to be hydrolysed by refluxing with concentrated sodium hydroxide solution?

a. Glyceryl tristearate
b. Propyl myristate
c. Sodium ricinoleate
d. Terylene
e. Nylon

18. Butyl acetate can be made by boiling equimolar quantities of butanol and acetic acid, the reaction being reversible. Which conditions will increase the percentage yield of the ester?

a. 2 moles of acid to 1 of alcohol.
b. 1 mole of acid to 2 of alcohol.
c. Remove the ester by distillation.
d. Remove water by distillation.
e. Add a mineral acid as a catalyst.

19. Compound A was treated with compound B under suitable conditions. The product, on boiling with excess sodium hydroxide solution, gave the sodium salt of an organic acid and liberated ammonia. After conversion of this salt to the parent acid and subsequent dehydration, the compound $C_4H_6O_2$ was produced which rapidly decolourised bromine water. A and B could have been:

	A	B
a.	CH_3CHO	NH_2OH
b.	CH_3COCH_3	$NaHSO_3$
c.	C_2H_5OH	HNO_3
d.	CH_3COCH_3	HCN
e.	CH_3CHO	NH_3

20. Which statements are correct concerning the industrial production and uses of these materials?

a. Carbon black is mainly used to make printing ink.
b. Formaldehyde can be made by the catalytic oxidation of methanol.
c. Formaldehyde is mainly used to preserve medical specimens.
d. Acetone is usually made by fermentation.
e. Acetic anhydride is used to acetylate cellulose.

AMINES, AMIDES AND CYANIDES

1. Degradation of an acid amide using bromine and warm sodium hydroxide solution gives ethylamine. The amide could be:

   a. $HCONH_2$
   b. CH_3CONH_2
   c. $CH_3CH_2CONH_2$
   d. $CH_3CH_2CH_2CONH_2$
   e. $CH_3CH_2CH_2CH_2CONH_2$

2. Propyl cyanide is reduced with hydrogen and a catalyst. The product is:

   a. Methylamine
   b. Ethylamine
   c. Propylamine
   d. Butylamine
   e. Butanoic acid

3. Catalytic reduction of an acid amide gives ethylamine. The amide is:

   a. $HCONH_2$
   b. CH_3CONH_2
   c. $CH_3CH_2CONH_2$
   d. $CH_3CH_2CH_2CONH_2$
   e. $CH_3CH_2CONHCH_3$

4. A white crystalline solid dissolves in water. On warming with excess sodium hydroxide solution, ethylamine is liberated. The compound could be:

   a. $CH_3CH_2NH_2.HCl$
   b. $(CH_3CH_2NH_3)^+ Cl^-$
   c. $CH_3CH_2NH_2.HNO_3$
   d. $(CH_3CH_2)_2NH.HCl$
   e. $CH_3CH_2CONH_2$

5. A white crystalline solid dissolves in water. Addition of excess cold acid and sodium nitrite solution gives a yellow oil. The compound could be:

   a. $(CH_3)_2NH$
   b. $(CH_3CH_2NH_2CH_3)^+ Cl^-$
   c. $(CH_3CH_2CH_2CH_2NH_3)^+ Cl^-$
   d. $[(CH_3CH_2)_3NH]^+Cl^-$
   e. $NH_2CH_2CH_2NH_2.2HCl$

6. A white crystalline solid is refluxed under a water-cooled condenser with excess sodium hydroxide solution. An organic base is left together with an aqueous solution which on evaporation of the water gives solid sodium acetate. The original solid could be:

   a. $C_6H_5NHCOCH_3$
   b. $CH_3NHCOCH_3$
   c. $CH_3CH_2CH_2NHCOCH_3$
   d. $(C_2H_5)_2CHNHCOCH_3$
   e. $(C_2H_5)_2CHNHCOCH_2CH_3$

AMINES, AMIDES AND CYANIDES

7. 1 mole of methylamine is heated under pressure with 3 moles of methyl iodide. The main organic product of this reaction is:

   a. $(CH_3NH_3)^+I^-$
   b. $(CH_3)_2NH$
   c. $(CH_3)_2NH_2.HI$
   d. $(CH_3)_3N$
   e. $[(CH_3)_4N]^+I^-$

8. In a laboratory preparation of acetamide 12·0 g of acetic acid were split into two equal parts. Ammonium carbonate solid was added to one portion until effervescence ceased. The other portion of acetic acid was then added and the mixture heated under an air-cooled condenser so that nothing but water distilled over. The acetamide so produced was finally separated from the excess acetic acid by distillation, and weighed 3·7 g.
   Which statements are correct?

   a. 0·1 mole of ammonium acetate is present.
   b. The ratio of moles of ammonium acetate to acetic acid is 2:1.
   c. The maximum weight of water which could be produced in the reaction is 1·8 g.
   d. 0·627 mole of acetamide is formed.
   e. The yield of acetamide is approximately 63 per cent.

9. In the method in Question 8 there are two main equilibria involved:
   i. ammonium acetate $\rightleftharpoons$ acetic acid + ammonia
   ii. ammonium acetate $\rightleftharpoons$ acetamide + water
   Which statements are correct?

   a. Excess acetic acid causes equilibrium **i** to move to the right.
   b. The removal of water causes equilibrium **ii** to move to the right.
   c. It is easier to remove acetamide from the mixture rather than water.
   d. A catalyst would move equilibrium **ii** to the right.
   e. Doubling the weight of ammonium acetate would increase the percentage yield of acetamide.

10. A compound on boiling with excess sodium hydroxide solution liberates ammonia.
    The compound could be:

    a. CH_3CONH_2
    b. $CH_3CH_2NH_2$
    c. CH_3COONH_4
    d. $CH_3CONHCH_3$
    e. CH_3CN

11. Ammonium carbonate solid is added to propanoic acid (a liquid) until the reaction ceases. The solid obtained after evapora-

AMINES, AMIDES AND CYANIDES

tion is heated with excess propanoic acid. The product from this is heated with phosphorus pentoxide, giving a liquid. This is:

a. Propanamide
b. Ethyl cyanide
c. Cyanoethane
d. Cyanopropane
e. Propionitrile

12. On addition of conc. ammonia solution to an organic liquid a crystalline solid is readily obtained. On refluxing with sodium hydroxide solution this gives ammonia. The original liquid could be:

a. $CHCl_2COOH$
b. CH_3COCl
c. $CH_3CH_2CH_2COCl$
d. $CH_3CH_2CH_2Cl$
e. $CH_3CH_2CH_2I$

13. Ethyl cyanide is refluxed for a long time with excess sodium hydroxide solution, liberating ammonia.
The final product left in solution is:

a. CH_3CH_2COOH
b. CH_3CH_2COONa
c. $CH_3CH_2COONH_4$
d. $CH_3CH_2CONH_2$
e. $CH_3CH_2NH_2$

14. An organic liquid is refluxed with alcoholic potassium cyanide solution, giving an alkyl cyanide.
The liquid could be:

a. CH_3OH
b. CH_3I
c. CH_3COCl
d. CH_3COOH
e. C_2H_5I

15. Methyl cyanide is refluxed with excess sodium and ethanol. The organic product is distilled into conc. hydrochloric acid, from which white crystals can be isolated. These crystals are:

a. Methylammonium chloride
b. Ethanamide
c. Ethylammonium chloride
d. Ammonium chloride
e. Acetic acid

16. An alcohol was refluxed with excess red phosphorus and iodine. The product was distilled and then refluxed with alcoholic potassium cyanide solution. The distillate from this was refluxed with excess sodium and ethanol. The organic product so formed was dissolved in hydrochloric acid and excess cold sodium nitrite solution

was added. From this final mixture was extracted propan-1-ol. The original alcohol was:

a. Methanol
b. Ethanol
c. Propan-2-ol
d. Methyl alcohol
e. Ethyl alcohol

17. An organic compound is treated with bromine and sodium hydroxide solution. The product is distilled off and dissolved in hydrochloric acid. Addition of excess sodium nitrite gives a mixture from which propan-1-ol can be extracted.
The original compound is:

a. $CH_3CH_2CONH_2$
b. $CH_3CH_2CH_2CONH_2$
c. $CH_3CH_2CONHCH_3$
d. $CH_3CH_2CON(CH_3)_2$
e. $CH_3CH_2COONH_4$

18. Indicate the properties you would expect this compound to have:
$NH_2CH_2CH_2COOH$

a. It will dissolve in hydrochloric acid forming Cl^- ($^+NH_3CH_2CH_2COOH$).
b. It will dissolve in sodium hydroxide solution forming $(NH_2CH_2CH_2COO)^-Na^+$.
c. In aqueous solution it will form the ion $^+NH_3CH_2CH_2COO^-$.
d. It will react with sodium nitrite and acid giving $HOCH_2CH_2COOH$.
e. It will effervesce with $NaHCO_3(aq)$.

19. A compound is boiled with excess sodium hydroxide solution for a few minutes, liberating ammonia. Acidification of the remaining solution enables crystals of an organic acid to be isolated, which in 0·1 M solution is a stronger acid than 0·1 M acetic acid.
The original solid could be:

a. $ClCH_2CONH_2$
b. $ClCH_2COONH_4$
c. Cl_3CCONH_2
d. $Cl_3CCOONH_4$
e. NH_2CH_2COCl

20. Which of these materials, on refluxing for a long time with excess sodium hydroxide solution, would you expect to give ammonia?

a. Nylon
b. Terylene
c. PVC
d. Courtelle (Polyacrylonitrile)
e. Rayon

AROMATIC COMPOUNDS

The benzene ring C_6H_6 is shown as

1. Which statements are correct concerning the structure of benzene?
   (The C—C bond length in ethane is 154 pm and in ethylene is 133 pm.)

   a. All the C and H atoms in the molecule are in the same plane.
   b. There are 3 stereoisomers of C_6H_5Cl.
   c. The C—C bond length has a value of 139 pm and is the same for all the bonds.
   d. The molecule contains fixed alternate double and single bonds.
   e. The ring exists in boat and chair forms.

2. Which statements are correct concerning the chemical reactions of benzene?

   a. 3 moles of bromine react additively with 1 mole of benzene in ultra-violet light.
   b. 1 mole of hydrogen reacts with 1 mole of benzene when the gas is passed into boiling benzene containing a catalyst, the reaction ceasing at this stage (i.e. at C_6H_8).
   c. Benzene is readily reduced by a dissolving metal reduction.
   d. Benzene can be oxidized appreciably by refluxing with conc. nitric acid.
   e. Iodobenzene is formed when iodine and benzene are refluxed together.

3. Which of these reagents will react with benzene to form substitution products?

   a. H_2SO_4(fuming)
   b. $ClSO_2OH$
   c. HCl(conc.)
   d. $KCN_{(aq)}$
   e. CH_3COCl (with $AlCl_3$ as a catalyst)

4. When 1 mole of benzene and 3 moles of chlorine react in ultra-violet light, a white solid results.
   Which statements are correct?

   a. This is an addition reaction.
   b. The molar mass of the solid is 184.
   c. In a molecule of the solid, all the atoms are in the same plane.
   d. The solid could be named hexachlorobenzene.
   e. The solid has little commercial use.

5. Chlorine was bubbled into 78 g of benzene under suitable conditions until the mixture weighed 113 g. Which statements are correct?

   a. The product will be mainly monochlorobenzene.
   b. The most likely impurity will be 1,3-dichlorobenzene.
   c. A catalyst is normally used.
   d. The product is a white solid.
   e. The reaction is exothermic.

6. Benzene (1 mole) is mixed with conc. nitric acid (1 mole) and conc. sulphuric acid (1 mole) with careful cooling until the reaction is complete. The mixture is then poured into water. What is true of the organic product?

   a. The product contains an appreciable proportion of benzene sulphonic acid.
   b. The product contains some 1,3-dinitrobenzene.
   c. The product is a colourless oil when purified.
   d. The introduction of one $-NO_2$ group into the ring makes the introduction of a second group easier.
   e. If the temperature is allowed to rise much above 60°C, then trinitrobenzene is readily formed.

7. Benzene is refluxed with excess conc. sulphuric acid.
   Which statements are correct concerning this reaction?

   a. Fuming sulphuric acid reacts faster.
   b. Unless care is taken a disubstituted product is readily formed.
   c. In the compound produced, the sulphur atom is not attached directly to the benzene ring.
   d. This type of reaction is of very little industrial significance.
   e. 0·02 mole of benzene sulphonic acid requires 20 ml M NaOH in a titration.

8. Acetyl chloride (1 mole) was added carefully to benzene (1 mole) containing aluminium chloride (0·3 mole). After refluxing, the mixture was poured into cold water.
   Which statements concerning the main organic product are correct?

   a. The product cannot be shaken with aqueous sodium carbonate as it readily reacts with alkalis.
   b. The product gives a precipitate with 2,4-dinitrophenylhydrazine when reacted under suitable conditions.
   c. If iodoethane were to be used instead of acetyl chloride, the product would then be ethylbenzene.

d. Aluminium chloride is the only catalyst which can be used for this type of reaction.
e. Oxidation of the product with acid potassium dichromate gives mainly acetic acid.

9. Chlorine is bubbled into boiling methylbenzene irradiated with ultra-violet light so that 1 mole of methylbenzene reacts with 1 mole of chlorine.
The product will be mainly:

a. C_6H_5Cl
b. $C_6H_5CH_2Cl$
c. C_6H_5COCl
d. $C_6H_5CHCl_2$
e. $ClC_6H_4CH_3$

10. A compound has these properties:
i. Only one of the chlorine atoms can be replaced by refluxing with aqueous sodium hydroxide.
ii. Only one of the chlorine atoms can be replaced by —CN when refluxed with alcoholic KCN.
The compound could be:

11. On prolonged refluxing with excess alkaline potassium permanganate, a compound was converted to benzoic acid. It could have been:

a. $C_6H_5CH_3$
b. $C_6H_5CCl_3$
c. Methylcyclohexane
d. $C_6H_5CH_2CH_3$
e. C_6H_5CHO

12. A compound, when dissolved in excess dilute acid and treated at about 5°C with a slight excess of sodium nitrite solution, is diazotised.
The original compound could be:

a. 1,4-Diaminobenzene

13. Aniline, when treated as in Question 12, gives a diazo compound in solution. When certain reagents are added to this solution a reaction occurs in which the final result is to replace —NH₂ in aniline by another group.
Which reagents work as suggested?

Reagent	Group replacing NH_2
a. C_2H_5OH	H
b. $K_3Cu(CN)_4$	CN
c. Warm water	OH
d. KI(aq)	I
e. HCl(dil) at 50°C	Cl

14. A compound has these properties:
   i. When mixed in aqueous solution with sodium hydrogen carbonate, there is no effervescence.
   ii. It dissolves in and reacts with aqueous sodium hydroxide.
   iii. It reacts readily with bromine water.
   The compound could be:

   a. C_6H_5OH
   b. $C_6H_5NH_2$
   c. $C_6H_5CH_2OH$
   d. $C_6H_5SO_2OH$
   e. C_6H_5COOH

15. Phenol is too weak an acid to liberate carbon dioxide from sodium carbonate solution.
Deduce which compounds are more likely to do this.

a. OH, NO₂ (on benzene ring)
b. OH, COOH (on benzene ring)
c. OH, NH₂ (on benzene ring)
d. OH, O₂N, NO₂, NO₂ (trinitrophenol on benzene ring)
e. OH, CH₃ (on benzene ring)

16. The dissociation constant for ammonia is 2×10^{-5} and for aniline is 5×10^{-10}. Deduce which of these compounds is most likely to have a dissociation constant of 1×10^{-14}.
(All values measured in water at 298 K.)

a. $C_6H_5CH_2NH_2$
b. $H_2NC_6H_4CH_3$
c. $H_2NCH_2C_6H_4CH_3$
d. $O_2NC_6H_4NH_2$
e. Aniline oxalate

17. Which of these compounds would you expect to react rapidly with bromine water, giving a substituted product?

a. Phthalic acid
b. 1,3-Dihydroxybenzene
c. Dinitrobenzene
d. N,N-Dimethylaniline
e. Styrene

18. Knowing the properties of acetone and acetic acid, deduce which compounds are most likely to give crystalline condensation products with both hydroxyammonia and 2,4-dinitrophenylhydrazine under suitable conditions.

a. $C_6H_5CONH_2$
b. $C_6H_5COC_6H_5$
c. C_6H_5CHO
d. C_6H_5COOH
e. $C_6H_5CONHC_6H_5$

19. Which of these compounds would you expect to possess aromatic characteristics?

a. Cyclohexanol
b. But-1,3-diene
c. Pyridine (C_5H_5N)
d. Benzenehexachloride
e. Cyclopentadienyl anion $(C_5H_5)^-$ as in ferrocene $(C_5H_5)_2Fe$

20. Which of these well known substances are made industrially from benzene or a similar aromatic hydrocarbon?

a. DDT
b. BHC
c. Nylon
d. Terylene
e. Polystyrene

AROMATIC AND ALIPHATIC COMPOUNDS

1. A compound is refluxed with excess sodium hydroxide solution for an hour. The aqueous layer is acidified with dilute nitric acid and when silver nitrate solution is added a thick precipitate forms. The compound could be:
   a. C_2H_5I
   b. C_6H_5Br
   c. $C_6H_5CH_2Br$
   d. $ClCH_2COOH$
   e. C_6H_5COCl

2. A mixture of two compounds is added to sodium hydrogen carbonate solution. Only one compound reacts, liberating carbon dioxide. The mixture could be:
   a. Phenol and benzoic acid
   b. Propanoic acid and $CH_3C_6H_4OH$
   c. Trichloroacetic acid and ethanol
   d. 2,4,6-Trinitrophenol and 1,3-dinitrobenzene
   e. Benzene sulphonic acid and $C_6H_5CH_2OH$

3. A compound when shaken with bromine water gives a white precipitate. The compound could be:
   a. $C_6H_5NH_2$
   b. C_6H_5COOH
   c. C_6H_5OH
   d. $NO_2C_6H_4OH$
   e. $C_6H_5SO_2O^-Na^+$

4. A compound is dissolved in dilute hydrochloric acid. On adding excess cold sodium nitrite solution and warming to room temperature a colourless, odourless gas is evolved. Which compound would you expect it to be?
   a. $C_6H_5CH_2NH_2$
   b. $C_4H_9NH_2$
   c. C_6H_5OH
   d. $O_2NC_6H_4OH$
   e. $C_6H_5NH_3^+Cl^-$

5. A compound liberates ammonia when refluxed with excess sodium hydroxide solution. The compound could be:
   a. Propanamide
   b. Ethylamine
   c. Aniline
   d. Urea
   e. Benzamide

6. A compound was dissolved in excess dilute hydrochloric acid. Excess sodium nitrite solution was added slowly at 5°C. On pouring the resulting solution into an alkaline solution of a second compound, a brightly coloured precipitate was produced. The two compounds could have been respectively:
   a. $C_6H_5CH_2NH_2$ and 2-naphthol
   b. 2-aminobenzoic acid and N,N-diethylaniline
   c. 4-aminobenzene sulphonic acid and N,N-dimethylaniline
   d. 2-methylaniline and 2-naphthol
   e. N,N-diethylaniline and benzoic acid

AROMATIC AND ALIPHATIC COMPOUNDS

7. A compound when shaken with bromine water and acidified potassium permanganate rapidly decolourised both of them. The compound could have been:
   a. Ethylene
   b. Benzene
   c. Cyclohexane
   d. Styrene
   e. Trichloroethane

8. A compound was refluxed with excess sodium hydroxide solution for an hour. On acidifying the resulting solution with dilute nitric acid, a white precipitate formed. This was removed by filtration. The filtrate gave another white precipitate when silver nitrate solution was added. The original compound could have been:
   a. $CH_3CH_2CH_2Cl$
   b. CH_3CH_2COCl
   c. C_6H_5Cl
   d. C_6H_5COCl
   e. $C_6H_5CCl_3$

9. A compound gives a coloured precipitate when heated with 2,4-dinitrophenylhydrazine in glacial acetic acid, the mixture then being poured into water. The original compound could be:
   a. $CH_3COCH_2CH_3$
   b. C_6H_5COCl
   c. CH_3COOH
   d. $C_6H_5CH_2CONH_2$
   e. CH_3CHO

10. A compound gives a precipitate as in Question 9, but also gives a red precipitate on boiling with Fehling's solution and a silver mirror on warming with ammoniacal silver nitrate.
    The compound could be:
    a. HCHO
    b. $C_6H_5CH_2CHO$
    c. $(CH_3CO)_2O$
    d. $C_6H_5CONH_2$
    e. CH_3CH_2CHO

11. A compound is boiled with excess sodium hydroxide solution, liberating ammonia. The compound could be:
    a. CH_3CN
    b. C_6H_5CN
    c. $C_6H_5CH_2NH_2$
    d. $CH_3CH_2CONH_2$
    e. $[(CH_3)_4N]^+Cl^-$

12. A compound forms a separate layer when shaken with water. On refluxing for an hour with excess sodium hydroxide solution the two layers merge to give a homogeneous mixture.
    The compound could be:
    a. Ethyl acetate
    b. Propyl propanoate
    c. Cetyl acetate
    d. Ethyl stearate
    e. Methyl benzoate

13. A liquid has two tests performed on separate samples:
    i. On warming with potassium iodide and sodium hypochlorite solutions it gives a yellow precipitate.
    ii. On oxidizing with acidified potassium dichromate solution, the volatile product gives an orange precipitate with 2,4-dinitrophenylhydrazine but has no effect on Fehling's or Tollens' reagents.
    The liquid could be:

    a. Methanol
    b. Ethanol
    c. Propan-1-ol
    d. Propan-2-ol
    e. Butan-1-ol

14. Two portions of a compound were treated separately:
    i. On refluxing with sodium hydroxide solution then acidifying with dilute nitric acid and filtering, the filtrate gave a precipitate with silver nitrate solution.
    ii. On adding to sodium hydrogen carbonate solution, effervescence occurred.
    The compound could have been:

    a. 3-chlorobenzoic acid (benzene ring with COOH and Cl)
    b. 2-bromophenol (benzene ring with OH and Br)
    c. CCl_3COOH
    d. CH_3CH_2COCl
    e. 1-chloro-4-nitrobenzene (benzene ring with Cl and NO_2)

15. 0·1 mole of a compound was refluxed with 100 ml of 5M alcoholic potassium hydroxide. The unused alkali required 40 ml of 5M hydrochloric acid in a titration. The compound could have been:

    a. $CH_3CH_2COOCH_2CH_3$
    b. $CH(COOCH_3)_2$ arrangement:
       COOCH$_3$ | COOCH$_3$
    c. benzene ring with COOCH$_3$ and COOCH$_3$
    d. benzene ring with O.CO.CH$_3$ and O.CO.CH$_3$
    e. $C_{17}H_{35}COOCH_2$ | $C_{17}H_{35}COOCH$ | $C_{17}H_{35}COOCH_2$

16. A compound was refluxed with excess tin and conc. hydrochloric acid. After making alkaline and steam distilling, the distillate was dissolved in dilute hydrochloric acid. This was treated with excess sodium nitrite

    a. $CH_3C_6H_4COOH$
    b. $CH_3C_6H_4NO_2$
    c. $CH_3C_6H_4NH_2$
    d. $CH_3CH_2C_6H_4NO_2$
    e. $O_2NC_6H_4NO_2$

solution at 5°C, and then warmed with ethanol when nitrogen was rapidly evolved. The oily liquid thus formed was separated and refluxed with acidified potassium dichromate solution, giving benzoic acid as the main product.
The original compound could have been:

17. A liquid is refluxed with excess potassium cyanide in ethanol. The product, after distillation, is reduced with sodium and ethanol to a basic compound. This dissolves in dilute hydrochloric acid and on adding sodium nitrite solution readily evolves an odourless gas. The original liquid could be:

   a. 1-Iodopropane
   b. 2-Iodopropane
   c. 1-Iodo-2-methylbutane
   d. C_6H_5I
   e. $C_6H_5CH_2I$

18. A compound rapidly decolourises bromine water and acidified potassium permanganate solution when shaken separately with them. When refluxed with excess sodium hydroxide solution, ammonia is evolved. The compound could be:

   a. C_2H_5CN
   b. $CH_2\!=\!CHCN$
   c. Acrylonitrile
   d. Acetamide
   e. C_6H_5CN

19. A compound is refluxed for some time with acidified potassium dichromate solution. From the products can be isolated an aromatic acid which forms an intramolecular anhydride on heating to 250°C. 0·01 mole of this anhydride requires 20 ml of M NaOH in a titration.
The original compound could be:

   a. benzene with CH_3 substituent
   b. benzene with CH_2CH_3 substituent
   c. benzene with NO_2 and CH_3 substituents
   d. benzene with two CH_3 substituents
   e. benzene with CH_3 and CH_2CH_3 substituents

20. Which of these metal compounds are of industrial and/or laboratory importance?

   a. $Al(C_2H_5)_3$
   b. CH_3MgI
   c. $Pb(C_2H_5)_4$
   d. $(CH_3COO)_2Ca$
   e. $LiAlH_4$

QUALITATIVE AND QUANTITATIVE ORGANIC ANALYSIS

1. An organic compound is heated with copper(II) oxide when it liberates carbon dioxide and water. A small amount on a copper wire turns a Bunsen flame green. The compound could be:

    a. C_2H_5I
    b. CCl_3CHO
    c. CHI_3
    d. C_2H_5OH
    e. CCl_3COOH

2. An organic compound is fused with sodium and the hot mixture is plunged into water. Sulphide and halide ion tests on this solution are negative, but CN^- is present. When the original compound is heated strongly for some time, an appreciable amount of inorganic residue remains. The compound could be:

    a. CH_3CONH_2
    b. CH_3COONH_4
    c. NH_2CH_2COONa
    d. CH_3CH_2CN
    e. $(CH_3CH_2CH_2NH_3)^+Cl^-$

3. A compound on a copper wire did not colour a Bunsen flame. It was fused with zinc dust and sodium carbonate, and the mixture was added to water. The residue effervesced with acid giving an odourless gas. Half of the filtrate gave a deep blue precipitate when an acidified mixture of iron(II) and iron(III) ions was added, and the other half gave a white precipitate when excess nitric acid and silver nitrate were added.
The solid could have been:

    a. CH_3CONH_2
    b. $C_6H_5SO_2NH_2$
    c. CH_3COCl
    d. CH_3CN
    e. $(CH_3NH_3)^+Cl^-$

4. Pure iodoethane must be separated from the mixture whose composition is given below.

Compound	Boiling point (°C)	% By weight
C_2H_5I	73	60
C_2H_5OH	78	10
H_2O	100	15
CH_3COOH	118	15

    An azeotrope exists of composition
      87% C_2H_5I/13% C_2H_5OH
         b.p. 63°C

Which of these methods would you expect to give a pure sample of iodoethane in reasonable yield?

a. Distil without a column and collect the fraction coming off first.
b. Distil with a simple column and collect at 73°C.
c. Reflux with excess sodium hydroxide solution to remove the acid, separate and distil the oily layer, and collect at 73°C.
d. Shake with excess sodium carbonate solution, separate the oily layer, dry with anhydrous sodium sulphate, distil and collect the first few drops.
e. Shake with excess sodium bicarbonate solution, separate the oily layer, dry it with anhydrous calcium chloride and distil (collecting at 73°C).

5. When crystallization is used to purify an organic compound, which points are important?

a. The compound should be very soluble in the solvent at all temperatures.
b. The solvent should be allowed to evaporate slowly at room temperature.
c. The compound should be soluble in the hot but insoluble in the cold solvent.
d. Impurities should remain dissolved in the cold mother liquor.
e. The process should be repeated until the melting point of the compound is constant.

6. Boiling points of saturated hydrocarbons are given. All have straight chains except one, which is branched. Plot a graph of boiling point against number of carbon atoms per molecule and decide which it is.

Hydrocarbon	Boiling point (°C)
Butane	0
Pentane	36
Hexane	49
Heptane	98
Octane	126
Nonane	151

a. Pentane
b. Hexane
c. Heptane
d. Octane
e. Nonane

7. 50 cm³ of a mixture of ethylene and acetylene were exploded with 160 cm³ of oxygen. After cooling, 100 cm³ of carbon dioxide were left in a total volume of 125 cm³ of gas. All volumes were measured at 298 K and $10^5 Nm^{-2}$.
Which statements are correct?

a. 3 moles of oxygen are used up by every mole of ethylene.
b. 3 moles of oxygen are used up by every mole of acetylene.
c. 1 mole of acetylene produces 2 moles of carbon dioxide.
d. The percentage by volume of ethylene is 40 per cent.
e. If ethane took the place of ethylene, the same experimental values would be obtained.

8. Which are isomers of heptane (C_7H_{16})?

a. 2-methylhexane
b. 3-methylhexane
c. 2,3-dimethylpentane
d. 2,4-dimethylpentane
e. 2,2,3-trimethylbutane

9. Which of these compounds has an isomer with a different structural or spatial configuration?

a.
```
    COOH
     |
  H—C—OH
     |
  H—C—OH
     |
    COOH
```

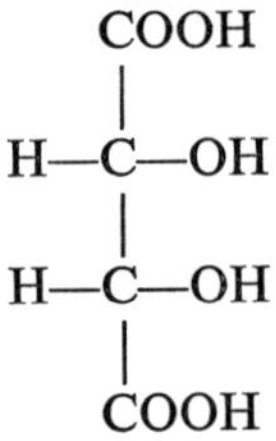

b.
```
H—C—COOH
   ||
H—C—COOH
```

c.
```
   CH₃
     \
      C=N—OH
     /
   H
```

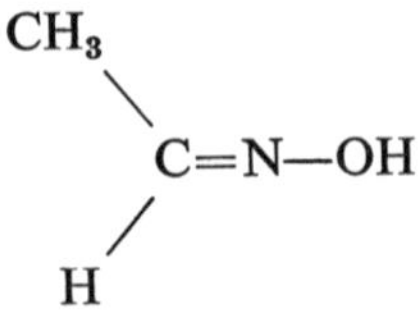

d.
```
   H  H
   |  |
H—C—C—Cl
   |  |
   H  Cl
```

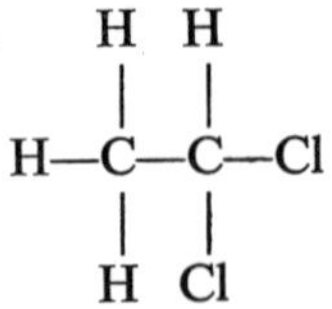

e.
```
   H  H
   |  |
H—C—C—OH
   |  |
   H  H
```

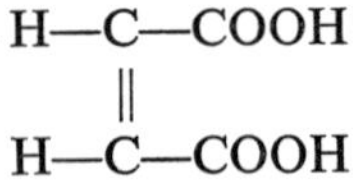

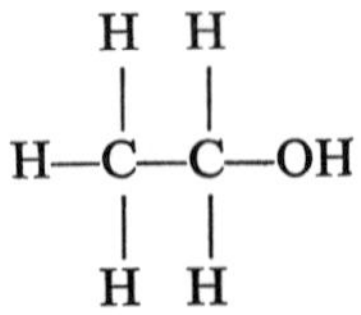

QUALITATIVE AND QUANTITATIVE ORGANIC ANALYSIS

10. Analysis gives the composition by weight of a compound as:
%C   52·2
%H   13·0
%O   34·8
1·0 of it gives with excess sodium 244 cm³ of hydrogen (collected dry and corrected to s.t.p.).
The compound could be:

a. Methanol
b. Ethanol
c. Propan-1-ol
d. Ethan-1,2-diol
e. Dimethyl ether

11. 1·02 g of a compound gave on complete oxidation in a suitable apparatus 1·84 g of carbon dioxide and 0·88 g of water. The same weight of another sample when decomposed in a Kjeldahl apparatus gave ammonia which was passed into 25·0 ml M HCl. The unused acid needed 11·0 ml M NaOH in a titration. The molar mass of the compound was found to be 74 ± 2. Which statements follow from these results?

a. 1 mole of compound gives 3 moles of carbon dioxide.
b. 1 mole of compound gives 3·5 moles of water.
c. 1 mole of compound gives 1 mole of ammonia.
d. The formula must be C_3H_7N.
e. The compound could be propanamide.

12. A compound was analysed by a method similar to that of Dumas. 1·15 g of compound gave 2·76 g of carbon dioxide, 0·94 g of water and 245 cm³ of nitrogen collected dry at 298 K and $10^5 Nm^{-2}$. The molar mass was 55 ± 5.
Which statements follow from these results?

a. 1 mole of compound gives 3 moles of carbon dioxide.
b. 1 mole of compound gives 0·5 mole of nitrogen.
c. The formula could be $C_2H_2N_2$.
d. The compound could be ethylene diamine.
e. The compound could be ethyl cyanide.

13. 0·01 mole of a compound is added to excess cold sodium hydroxide solution. After shaking for a few minutes the solution is acidified with nitric acid. From this 1·43 g of silver chloride can be obtained. The compound could be:

a. Monochloroacetic acid
b. Dichloroacetic acid
c. Acetyl chloride
d. Ethyl chloride
e. The acyl chloride of any monobasic acid.

QUALITATIVE AND QUANTITATIVE ORGANIC ANALYSIS

14. 0·335 g of a compound was completely burned in oxygen in a flask and the products absorbed in alkaline hydrogen peroxide solution. After boiling to decompose unused peroxide, and acidifying with nitric acid, the halide produced required 25·5 ml 0·1M $AgNO_3$ to reach the end point when titrated potentiometrically.
Which statements follow from these results?

   a. 1 mole of $AgNO_3$ is needed to titrate all the halide from 131 g of compound.
   b. The molar mass of the compound must be at least 131.
   c. The compound could be any dibromoalkane with a molar mass of 262.
   d. The compound might be CHI_3.
   e. The compound might be monochloroacetic acid.

15. 0·88 g of a compound which has no effect on indicator paper is refluxed for two hours with 25·0 ml M NaOH. The excess alkali requires 15·0 ml M HCl in a titration.
The empirical formula of the compound is C_2H_4O. It could be:

   a. Acetaldehyde
   b. Propyl formate
   c. Ethyl acetate
   d. Methyl propanoate
   e. Butanoic acid

16. 0·01 mole of a solid organic acid was boiled with excess aqueous ammonia until no more ammonia gas was liberated. Excess silver nitrate solution was then added and the resulting precipitate washed, dried and then strongly heated in a crucible to constant weight. The residue so obtained weighed 2·16 g. The acid could have been:

   a.
   $$\begin{array}{c} COOH \\ | \\ H-C-OH \\ | \\ H-C-OH \\ | \\ COOH \end{array}$$

   b. $C_2H_2O_4, 2H_2O$
   c. Oxalic acid
   d. Citric acid (2-hydroxypropane-1,2,3-tricarboxylic acid)
   e. Any solid dibasic organic acid

17. A pure oil isolated from linseed oil has a molecular weight of 880. By a standard procedure the Iodine Number was found to be 170. (This is the number of grammes of iodine combining with 100 g of oil.)
Which statements are correct?

a. 6 moles of iodine (I_2) are taken up by 1 mole of oil.
b. There are three double bonds present in a molecule of oil.
c. The compound could be the triglyceride of octadec-9,12-dienoic acid.
d. The minimum quantity of sodium hydroxide needed to completely saponify 1 mole of the oil is probably 2 moles.
e. The oil would be expected to harden on exposure to air.

18. Compound A was dissolved in dilute hydrochloric acid and treated with excess sodium nitrite solution, giving B.
B was refluxed with red phosphorus and iodine (both in excess) to give C.
C was refluxed with excess potassium cyanide solution to give D.
D, when boiled with excess sodium hydroxide solution, liberated ammonia and left the disodium salt of the dibasic acid $C_3H_4O_4$.
Compound A could have been:

a. Acetamide
b. Methylammonium chloride
c. Ammonium acetate
d. Aminoacetic acid
e. Cyanoethane

19. Compound A contains nitrogen. After refluxing with excess sodium hydroxide it gives B which contains no nitrogen.
B combines with hydrogen cyanide (in the presence of potassium cyanide) to give C.
C is refluxed for some time with dilute hydrochloric acid to give D.
D can be dehydrated to E which rapidly decolourises bromine in carbon tetrachloride thus giving F.
F has the molecular formula $C_3H_4Br_2O_2$.
Compound A could be:

a. Propanamide
b. Ethylammonium chloride
c. Urea
d. Acetaldoxime
e. Glycine

20. Compound A when boiled with excess sodium hydroxide solution liberated ammonia. Acidification of the remaining solution gave B.

B was treated under pressure with hydrogen and a nickel catalyst until no more gas was taken up. This gave C.

1 mole of C was refluxed with red phosphorus while chlorine was bubbled in until there was a weight increase of 34·5 g. This gave D.

D was refluxed with sodium hydroxide solution and on acidification E was formed. E has the molecular formula $C_3H_6O_3$ and could be resolved into two optically active forms.

Compound A could have been:

a. $HCONH_2$
b. C_2H_5CN
c. $CH_2{=}CHCN$
d. $(CH_3)_2C(OH)CN$
e. $C_2H_4(NH_2)_2$

PHYSICAL REVISION 1

Only one answer is correct.

1. 2·0 g of a monobasic acid was made up to 250 ml with distilled water. 25 ml of this solution reacted completely with 20 ml 0·1 M NaOH. A mole of acid weighs
   a. 100 g
   b. 50 g
   c. 200 g
   d. 32 g
   e. 64 g

2. Aqueous solutions of two different monobasic organic acids are found to have the same degree of dissociation into ions (α) at 25°C. Which of the following statements is true?
   a. They must be the same molarity.
   b. They must have the same pH value.
   c. They must have the same dissociation constant.
   d. They must have the same hydrogen ion concentration.
   e. We cannot say which is the stronger unless we know the concentration.

3. A 0·001 M solution of sodium hydroxide has a pH value of
   a. 10^3
   b. 3
   c. 10^{-3}
   d. 11
   e. 10^{-11}

4. Which of the following changes is an oxidation?
   a. Hydroxylamine to ammonia.
   b. NO_2^- ions to nitric oxide gas.
   c. H^+ ions to hydrogen gas.
   d. Ammonium ions to ammonia gas.
   e. Hydrazine to nitrogen.

5. Which of the following solutions require 25 ml of 0·02 M $KMnO_4$ (in the presence of excess dilute sulphuric acid) for complete oxidation?
   a. 50 ml 0·1 M $FeSO_4$.
   b. 25 ml 0·02 M oxalic acid.
   c. 25 ml 0·01 M hydrogen peroxide.
   d. 25 ml 0·05 M sodium nitrite.
   e. 25 ml 0·1 M sodium oxalate.

6. Solid sodium acetate is added to a molar solution of acetic acid. Which of the following statements is true?
   a. The hydrogen ion concentration decreases.
   b. The acetic acid becomes stronger.
   c. The acetic acid solution becomes more concentrated.

d. The dissociation constant of acetic acid increases.
e. More acetic acid ionizes.

7. Water is added to M propanoic acid. Which of the following statements is true?
   a. [H⁺] is decreased.
   b. pH is decreased.
   c. The acid becomes weaker.
   d. The degree of dissociation (α) decreases.
   e. The dissociation constant decreases.

8. According to the Brønsted-Lowry concept, an acid is a proton donor and a base is a proton acceptor. Which of the following statements is FALSE?
   a. When ammonia is added to water the water acts as an acid.
   b. When hydrogen chloride is added to water the water acts as a base.
   c. When concentrated sulphuric acid is added to concentrated nitric acid, the nitric acid acts as a base.
   d. Ammonia is only a base in the presence of water.
   e. Ammonium chloride acts as an acid in liquid ammonia.

9. Which of the following is a necessary condition when using any of the colligative properties of solutions as a method of determining the molecular weight of a given solute?
   a. A non-volatile solvent.
   b. A solute of negligible vapour density.
   c. A solvent of negligible vapour pressure.
   d. A dilute solution.
   e. An aqueous solution.

10. 0·57 g of a weak monobasic organic acid reacts with 25 ml 0·1 M NaOH. The freezing point of benzene is lowered by 0·512°C when 4·56 g of the acid are dissolved in 100 g of benzene. The freezing point depression of benzene is 5·12 K mol⁻¹ kg.
    Which of the following is true?
    a. The molar mass by titration is 22·8.
    b. The molar mass by freezing point is 45·6.
    c. The acid molecule contains two replaceable hydrogen atoms.
    d. The acid is completely dissociated into ions in the water.
    e. The acid is dimerised in the benzene.

11. Which of the following statements about methylamine (CH_3NH_2) do you think is FALSE?
    a. It is a substituted ammonia.
    b. It is a strong base.
    c. It is a gas at room temperature and is very soluble in water.
    d. It forms white smoke with hydrogen chloride gas.
    e. It forms a complex ion with Cu^{2+}.

12. The solubility products of some of the sparingly soluble salts of silver are quoted as:

    AgCl 2×10^{-10}
    AgBr 5×10^{-13}
    Ag_2CO_3 8×10^{-12}
    Ag_2S 6×10^{-50}
    AgI 8×10^{-17}

    Which of these salts is the most soluble, expressed in mol l^{-1}?

    a. AgCl
    b. AgBr
    c. Ag_2CO_3
    d. Ag_2S
    e. AgI

13. Given the information in Question 12 and the stability constants for the reactions:
    $Ag^+ + 2CN^- \rightleftharpoons Ag(CN)_2^-$:$K = 10^{21}$
    $Ag^+ + 2NH_3 \rightleftharpoons Ag(NH_3)_2^+$:$K = 10^7$
    Which of the following statements do you think is most likely to be true?

    a. Silver sulphide dissolves in excess M NH_3(aq).
    b. Silver iodide dissolves in excess M NH_3(aq).
    c. Silver sulphide dissolves in excess M KCN(aq).
    d. Silver bromide dissolves in excess M NH_3(aq).
    e. Silver carbonate is insoluble in M NH_3(aq).

14. Over which solution will there be the greatest pressure of water vapour?

    a. M sodium chloride (NaCl) solution at 20°C.
    b. M glucose ($C_6H_{12}O_6$) solution at 50°C.
    c. M urea (CON_2H_4) solution at 20°C.
    d. M calcium chloride ($CaCl_2$) solution at 20°C.
    e. M aluminium sulphate ($Al_2(SO_4)_3$) solution at 50°C.

15. The specific resistance of 0·01 M potassium chloride is found to be 8000 ohm cm^{-1}. The molar conductivity of potassium chloride in ohm^{-1} cm^2 mol^{-1} is:

    a. 800 000
    b. 80
    c. 12·5
    d. $1·25 \times 10^{-2}$
    e. $1·25 \times 10^{-4}$

16. Molar potassium cyanide was added slowly to 100 ml of a 0·1 M solution of iron (II) sulphate in which dipped two platinum electrodes. The resistance of the solution was found by including the cell

    a. The resistance was directly proportional to the volume of potassium cyanide added.
    b. The resistance was inversely proportional to the volume of potassium cyanide added.
    c. The resistance decreased slowly until 60 ml

in a Wheatstone bridge circuit (a.c. pattern). Which of the following do you think is most likely to be a correct statement of what happened as the potassium cyanide was slowly added?

had been added, and then decreased rapidly.
d. The resistance increased slowly until 40 ml had been added, and then decreased rapidly.
e. The resistance decreased slowly until 10 ml had been added, and then decreased more rapidly.

17. The Daniell primary cell is usually represented as
$$Zn \mid Zn^{2+} \parallel Cu^{2+} \mid Cu$$
The e.m.f. of the cell is usually given as 1·1 volts. Which of the following would be a possible method of increasing this e.m.f.?

a. Using concentrated zinc sulphate solution.
b. Using very dilute zinc sulphate solution.
c. Using M sulphuric acid instead of copper sulphate.
d. Using large electrodes.
e. Using cadmium in an M solution of cadmium sulphate instead of zinc in M zinc sulphate.

18. In some textbooks potash alum is given as $K_2SO_4, Al_2(SO_4)_3, 24H_2O$ and in others as $KAl(SO_4)_3, 12H_2O$.
Which of the following statements is true?

a. On heating, the former molecule dissociates into two molecules of the latter.
b. One is the formula gained by finding the molecular weight by colligative properties, whilst the other is found by some other method (e.g. vapour density).
c. The second formula is that gained from finding the molecular weight in aqueous solution, and the first formula in some other solvent in which dimerisation took place.
d. Either formula is permissible but the second formula is the result of analysis and the first shows the ratio of the weights of K_2SO_4 to $Al_2(SO_4)_3$.
e. The first formula is in the crystal form and the second is in aqueous solution.

19. Which of the following statements would you consider to be FALSE?

a. Tin and lead form a eutectic alloy.
b. Copper and nickel form solid solutions.
c. Rhodium and ruthenium form solid solutions.
d. Cadmium and zinc form solid solutions.
e. Magnesium and zinc form a compound.

20. Water and hydrogen chloride form an azeotrope (20·2 per cent HCl by weight) which boils at 108·6°C.
Which of the following statements do you consider to be FALSE?

a. 10 per cent hydrochloric acid becomes more concentrated on boiling.
b. 30 per cent hydrochloric acid becomes less concentrated on boiling.
c. 10 per cent hydrochloric acid boils at 104°C.
d. 40 per cent hydrochloric acid boils at 112°C.
e. The constant boiling mixture can be used in the preparation of a standard acid for volumetric analysis.

PHYSICAL REVISION 2

Only one answer is correct.

1. Which of the following require the greatest volume of M NaOH for complete reaction?
   a. 45 ml M H_2SO_4
   b. 75 ml M HCl
   c. 50 ml M oxalic acid
   d. 80 ml M formic acid
   e. 90 ml M acetic acid

2. Which of the following would react with the greatest volume of M NaOH solution?
   a. 0·1 mole of sulphuric acid
   b. 0·1 mole of copper sulphate
   c. 0·1 mole of phenylammonium chloride (aniline hydrochloride)
   d. 0·1 mole of zinc sulphate
   e. 0·1 mole of benzoic acid

3. Bromocresol green is an indicator which is a weak acid. The dissociation constant is 2×10^{-5}.
   At what pH would you expect it to change colour?
   a. 2·5
   b. 4·2
   c. 4·7
   d. 5·2
   e. 5·3

4. A buffer consisting of a mixture of M acetic acid and M sodium acetate is to be made with pH 3·5.
   In what proportions would you mix the solutions?
   ($pK_{acetic\ acid} = 4·8$)

	acid		salt
a.	13 ml	:	10 ml
b.	10 ml	:	13 ml
c.	20 ml	:	10 ml
d.	20 ml	:	1 ml
e.	10 ml	:	1 ml

5. Which of the following changes requires the presence of an oxidizing agent such as hydrogen peroxide for the reaction to take place?
   a. Dissolve aluminium oxide in sodium hydroxide solution.
   b. Dissolve zinc oxide in sodium hydroxide solution.
   c. Dissolve chromium(III) oxide in sodium hydroxide.
   d. Dissolve copper in dilute nitric acid.
   e. Convert potassium chromate into potassium dichromate.

6. Which of the following changes will NOT take place in the presence of a large concentration of H+ ions?
   a. Reduction of MnO_4^- to Mn^{2+} by Fe^{2+} ions.
   b. Reduction of $Cr_2O_7^{2-}$ to $2Cr^{3+}$ by ethanol.
   c. Oxidation of I^- to $\frac{1}{2}I_2$ with H_2O_2.
   d. Precipitation of CuS from aqueous $CuSO_4$ with H_2S.
   e. Precipitation of ZnS from aqueous $ZnSO_4$ with H_2S.

7. Which oxidation states of titanium are there amongst the compounds shown below?
   TiO_2, Na_2TiO_3, $Na_2Ti_2O_5$, K_2TiF_6, $(NH_4)_2TiBr_6$, K_2TiCl_5.
   a. 2, 3 and 4
   b. 2, 3
   c. 3, 4
   d. 2 only
   e. 4 only

8. The following gases were contained in a closed steel vessel of 22·4 dm³ capacity at 0°C:
   0·4 g of hydrogen
   2·8 g of nitrogen
   0·8 g of helium
   What is the pressure of the gas in the vessel?
   a. 0·35 atmospheres
   b. 0·40 atmospheres
   c. 380 mmHg
   d. 532 mmHg
   e. 10^5 Nm^{-2}

9. In which of the following reactions would a big increase in pressure be likely to bring about the largest increase in the products of the right-hand side of the equation?
   a. $2SO_2 + O_2 \rightleftharpoons 2SO_3$
   b. $SO_2 + NO_2 \rightleftharpoons NO + SO_3$
   c. $CO + 2H_2 \rightleftharpoons CH_3OH$
   d. $C_2H_4 + H_2 \rightleftharpoons C_2H_6$
   e. $N_2 + 3H_2 \rightleftharpoons 2NH_3$

10. Which of the following would boil at the lowest temperature?
    a. A solution of 0·585 g of NaCl in 100 g of water.
    b. A solution of 3·0 g of urea in 100 g of water.
    c. A solution of 4·4 g of ethanol in 100 g of water.
    d. A solution of 0·01 mole of $CaCl_2$ in 100 g of water.
    e. M NaOH.

11. Solid calcium hydroxide was shaken with 0·01 M NaOH, and the solid allowed to settle. 100 ml of the supernatant liquid required 30 ml of 0·1 M HCl for neutralisation. The solubility product of $Ca(OH)_2$ is:
    a. 1×10^{-6}
    b. 3×10^{-6}
    c. 4×10^{-6}
    d. 6×10^{-6}
    e. 9×10^{-6}

12. Dilute hydrochloric acid is added to separate solutions of Mn^{2+}, Cd^{2+}, Ni^{2+}, and Co^{2+} until the pH becomes 3. H_2S is now bubbled into these solutions. Assume that the concentrations of the metal ions are molar and that in a saturated solution of H_2S $[H^+]^2[S^{2-}] = 10^{-23}$.
Solubility products: MnS 10^{-15}
   CdS 10^{-28}
   NiS 10^{-24}
   CoS 10^{-26}
Which metal sulphides should be precipitated?

a. CdS only
b. MnS, NiS, and CoS
c. CdS and CoS
d. CdS, CoS, and NiS
e. None of them

13. M HCl is run into 0·1 M NaOH from a burette. Which of the suggested methods of finding the end point of the titration do you think would be of *least* practical value?

a. Change in colour of an indicator.
b. The pH falls rapidly.
c. The temperature ceases to rise.
d. The rate of change of specific gravity of the solution alters.
e. The conductance of the solution rises sharply.

14. Iodine dissolves in a variety of solvents to give either a brown (water or ethanol) or a violet (CCl_4) solution. If you were asked to investigate whether iodine existed in different molecular forms in the two solvents, which do you consider would be the best line of approach?

a. Find the freezing point depression of two solutions of the same molarity.
b. Find the boiling point elevation of the same concentration of iodine in the two solvents.
c. Find the heat of solution in each solvent.
d. If the two solvents are immiscible find the distribution ratio between the two solvents.
e. Find the conductance of two solutions of the same molarity.

15. The standard e.m.f. values of the following cells are as given.
H_2 | $2H^+$ || Cu^{2+} | Cu $E° = 0·34$ volt
Mg | Mg^{2+} || Cd^{2+} | Cd $E° = 1·94$ volt
Cd | Cd^{2+} || Cu^{2+} | Cu $E° = 0·74$ volt
Mg | Mg^{2+} || Ni^{2+} | Ni $E° = 2·09$ volt
What is the standard electrode potential for the reaction $Ni^{2+} + 2e^- \rightleftharpoons Ni$?

a. $+1·23$ volt
b. $+0·93$ volt
c. $-0·25$ volt
d. $-0·55$ volt
e. $-0·93$ volt

16. 0·108 g of silver is deposited by the passage of 0·1 amp for 16 minutes during the electrolysis of aqueous silver nitrate. If we accept that the Avogadro constant is 6×10^{23}, what is the approximate charge on the electron? (Ag = 108.)

a. $1·6 \times 10^{-19}$ coulomb
b. $6·0 \times 10^{-19}$ coulomb
c. $2·67 \times 10^{-22}$ coulomb
d. $2·67 \times 10^{-24}$ coulomb
e. $6·0 \times 10^{-27}$ coulomb

17. Given the following facts:
$Ag(s) \rightarrow Ag(g)$: $\Delta H = + 280$ kJ
$Ag(g) \rightarrow Ag^+(g)$: $\Delta H = + 732$ kJ
$Ag^+(g) \rightarrow Ag^+(aq)$: $\Delta H = - 473$ kJ
$\frac{1}{2}H_2(g) \rightarrow H^+(aq)$: $\Delta H = + 452$ kJ
Calculate the e.m.f. of the standard cell:
$\frac{1}{2}H_2 \mid H^+ \parallel Ag^+ \mid Ag$
Charge on the electron = $1·6 \times 10^{-19}$ C
Avogadro constant = 6×10^{23} mol^{-1}

a. 0·80 volt
b. 0·87 volt
c. 0·91 volt
d. 1·03 volt
e. 10·3 volt

18. *Ionic radii in* pm
Cs$^+$ 169
K$^+$ 133
Na$^+$ 95
Cl$^-$ 181
I$^-$ 216
Which of the following statements would you think was FALSE?

a. A NaCl crystal consists of interpenetrating face centred cubic structures.
b. Each Na$^+$ ion and each Cl$^-$ ion in NaCl has six nearest neighbours of opposite sign and twelve nearest neighbours of the same sign.
c. The CsCl crystal consists of a body centred cubic structure in which the two different ions occupy alternate positions.
d. In CsCl each ion has eight nearest neighbours of opposite sign.
e. KI would be expected to have the same structure as CsCl.

19. Silver (atomic mass 108) crystallises with a face-centred lattice. The density of silver is $1·08 \times 10^4$ kg m^{-3}. If we accept the Avogadro constant to be 6×10^{23} mol^{-1}, which is the FALSE step in the following calculation to find the radius (R) of the silver atom?

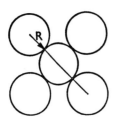

a. Volume of unit cube of silver is
$$\left(\frac{4R}{\sqrt{2}}\right)^3$$
b. There are the equivalent of four silver atoms in the unit cube.
c. There are L/4 unit cubes in 10^{-5} m^3 of silver.
d. $\dfrac{L}{4}\left(\dfrac{4R}{\sqrt{2}}\right)^3 = 10^{-5}$ m^3 mol^{-1}.
e. $R = 2·8 \times 10^{-10}$ m.

20. In the Downs cell for the production of sodium, a molten mixture of 40 per cent NaCl and 60 per cent CaCl$_2$ is electrolysed at 580°C.
Which statement is FALSE concerning this process?

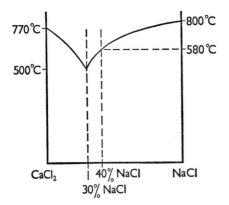

a. Some calcium is produced in the cell.
b. A melt of 30 per cent NaCl freezes suddenly at 500°C.
c. A solid mixture of 40 per cent NaCl melts suddenly at 580°C.
d. CaCl$_2$ and NaCl do not form a solid solution.
e. When a melt of 40 per cent NaCl is cooled, sodium chloride is the first solid to form as crystals.

PHYSICAL REVISION 3

Only one answer is correct.

1. Avogadro's Law was of great value in the development of chemical theory in the last century.
   Which of these facts had no historical connection with the evolution of the Law?

   a. Hydrogen is diatomic.
   b. Helium is monatomic.
   c. The molecular weight of a gas is twice the vapour density.
   d. 1 mole of gas occupies 22·4 litres at s.t.p.
   e. The atomic weight of carbon is 12.

2. Given only that the formula of lithium selenide is Li_2Se, which of the following statements do you KNOW to be true? (As distinct from those statements which may be true, but are nevertheless a matter of conjecture.)

   a. A molecule of Li_2Se contains two atoms of lithium and one atom of selenium.
   b. The solution of Li_2Se in water contains twice as many Li^+ ions as Se^{2-} ions.
   c. In any sample of Li_2Se there are twice as many atoms of lithium as there are selenium atoms.
   d. Molten lithium selenide contains twice as many Li^+ ions as Se^{2-} ions.
   e. The crystal of lithium selenide consists of a lattice of Li^+ and Se^{2-} ions.

3. Which of the following methods would be the most suitable to determine the molecular weight of a new polymer which was found to be soluble in chloroform ($CHCl_3$)?

   a. Determination of the vapour density by Victor Meyer's method.
   b. Elevation of the boiling point of chloroform.
   c. Depression of the freezing point of chloroform.
   d. Lowering of the vapour pressure of chloroform.
   e. Determination of the osmotic pressure of a solution in chloroform.

4. If you were asked to provide a suitable liquid to demonstrate Victor Meyer's method of molecular weight determination, which of the following would you choose? (You would not want to get involved in a discussion about thermal dissociation.)

   a. Methylamine CH_3NH_2
   b. Formaldehyde (HCHO)
   c. Methyl chloride (CH_3Cl)
   d. Trichlorethylene ($CHCl=CCl_2$)
   e. Trichloroacetic acid (CCl_3COOH)

5. Which of the following, when 1 g is dissolved in 100 g of water, would lower the freezing point by the greatest amount?

a. Common salt
b. Ethanol
c. Glycol (antifreeze)
d. Urea (CON_2H_4)
e. Calcium chloride

6. Calculate the vapour pressure at 25°C of an aqueous solution of glycerol $(C_3H_8O_3)$ containing 46 g in 253 g of solution. The vapour pressure of water at 25°C can be taken as $3 \cdot 12$ kN m^{-2}.

a. $1 \cdot 56$ kN m^{-2}
b. $2 \cdot 00$ kN m^{-2}
c. $2 \cdot 38$ kN m^{-2}
d. $2 \cdot 60$ kN m^{-2}
e. $2 \cdot 99$ kN m^{-2}

7. $2 \cdot 8$ g of an ammonia derivative of cobalt chloride lowered the freezing point of 100 g of water by $0 \cdot 62$°C. Which of the following compounds agrees with this information? (Molecular depression is $18 \cdot 6$°C for 100 g of water.)

a. $Co(NH_3)_3Cl_3$
b. $[Co(NH_3)_4Cl_2]^+Cl^-$
c. $[Co(NH_3)_5Cl]^{2+}2Cl^-$
d. $[Co(NH_3)_5H_2O]^{3+}3Cl^-$
e. $[Co(NH_3)_6]^{3+}3Cl^-$

8. 1 mole of nitrogen gas and 3 moles of hydrogen gas were contained in a steel vessel at 400°C and 400 atmospheres pressure.
At equilibrium it was found that there was 54 per cent NH_3 by volume in the equilibrium mixture.
Which of the following alterations in the conditions of temperature and pressure would be the most likely to produce the stated percentage of NH_3 in the mixture at equilibrium.
$N_2 + 3H_2 \rightleftharpoons 2NH_3 : \Delta H = -46 \cdot 2$ kJ

a. At 200°C and 200 atmospheres there is 90 per cent NH_3 in the mixture.
b. At 200°C and 1000 atmospheres there is 99 per cent NH_3.
c. At 500°C and 500 atmospheres there is 15 per cent NH_3.
d. At 500°C and 1000 atmospheres there is 20 per cent NH_3.
e. At 500°C and 200 atmospheres there is 60 per cent NH_3.

9. The equilibrium constant for the reaction
$H_2 + I_2 \rightleftharpoons 2HI$
is given as 49 at 400°C.
If equal volumes of hydrogen and iodine are heated together in a closed vessel at 400°C, what fraction by volume of the equilibrium mixture is hydrogen iodide?

a. 7/9
b. 7/8
c. 8/9
d. 4/9
e. 4/11

10. Given the following dissociation constants:

   formic acid $\quad 10^{-3.8}$
   acetic acid $\quad 10^{-4.8}$
   propanoic acid $\quad 10^{-4.9}$
   monochloroacetic acid $10^{-2.9}$
   dichloroacetic acid $\quad 10^{-1.3}$
   trichloroacetic acid $\quad 10^{-0.7}$

   which of the following statements do you think is FALSE?

   a. The strongest acid quoted is trichloroacetic acid.
   b. The introduction of electron acceptor groups into the acetic acid molecule makes the acid stronger.
   c. Formic acid is a stronger acid than propanoic acid.
   d. K_a for monobromoacetic acid is $10^{-2.3}$.
   e. K_a for butanoic acid is $10^{-4.9}$.

11. Given that pK_a for propanoic acid is 4·9 calculate the pH of a 0·01 M solution.

   a. 2·5
   b. 3·0
   c. 3·5
   d. 4·0
   e. 4·5

12. Methyl orange is a weak acid of dissociation constant 2×10^{-4}. At what pH would you expect it to change colour if we assume that at the point of change the concentration of the anions is the same as that of the free acid.

   a. 3·7
   b. 4·3
   c. 8·0
   d. 9·7
   e. 10·3

13. 1 mole of solid NH_4Cl was added to 1 litre of 0·1 M ammonia solution, and this solution was added to separate molar solutions of $FeCl_3$, $Cr_2(SO_4)_3$, $FeSO_4$, and $MgCl_2$. Given the solubility products $Fe(OH)_2 = 10^{-16}$, $Fe(OH)_3 = 10^{-38}$ $Mg(OH)_2 = 10^{-11}$, $Cr(OH)_3 = 10^{-29}$ and the dissociation constant for aqueous ammonia
   $K_b = 1·6 \times 10^{-5}$
   which of the following statements is true?

   a. All of the hydroxides should be precipitated.
   b. None of the hydroxides is precipitated.
   c. One would expect only $Fe(OH)_3$ to be precipitated.
   d. One would expect all except $Mg(OH)_2$ to be precipitated.
   e. One would expect $Fe(OH)_2$ and $Fe(OH)_3$ only to be precipitated.

14. The following constitute different types of reversible reactions in which a position of equilibrium exists.
   i. Dissociation of an acid
   $$C_6H_5COOH \rightleftharpoons C_6H_5COO^- + H^+$$

ii. Saturated solution of a sparingly soluble salt
(CuS(solid) $\rightleftharpoons Cu^{2+} + S^{2-}$)
iii. Formation of a complex ion
$Cu^{2+} + 4NH_3 \rightleftharpoons Cu(NH_3)_4^{2+}$
iv. $Fe^{3+} + 6CN^- \rightleftharpoons Fe(CN)_6^{3-}$
v. Formation of an ester
$CH_3COOH + C_2H_5OH$
$\rightleftharpoons CH_3COOC_2H_5 + H_2O$
Which statement do you think is most likely to be true?

a. Dissociation constant (K_a) in i. is 10^{24}.
b. Solubility product (S) in ii. is 10^{45}.
c. Stability constant in iii. is 0·3.
d. Stability constant in iv. is 10^{31}.
e. Equilibrium constant in v. is 10^{25}.

15. Which oxidation states of molybdenum are there among the compounds shown below?
$K_4Mo(CN)_8$, $(NH_4)_2MoO_4$, MoO_3, Mo_2O_5, K_2MoF_8, K_3MoCl_6.

a. 2, 3, 4, 5, 6
b. 2, 3, 4, 6
c. 2, 4, 6
d. 2, 4, 5, 6
e. 3, 4, 5, 6

16. Given that
$Zn^{2+} + 2e^- \rightleftharpoons Zn$ $E^\circ = -0.76$ volt
$Cu^{2+} + 2e^- \rightleftharpoons Cu$ $E^\circ = +0.34$ volt
which of the following statements do you think is *unreasonable*?
(N.B. If the left-hand electrode is the negative of the cell then E is positive.)

a. $Zn \mid M\ Zn^{2+} \parallel M\ Cu^{2+} \mid Cu$ $E^\circ = 1.1$ volt
b. $Zn \mid \dfrac{M}{100}\ Zn^{2+} \parallel \dfrac{M}{100}\ Cu^{2+} \mid Cu$ $E^\circ = 1.1$ volt
c. $Zn \mid \dfrac{M}{100}\ Zn^{2+} \parallel M\ Cu^{2+} \mid Cu$ $E^\circ = 1.2$ volt
d. $Zn \mid M\ Zn^{2+} \parallel \dfrac{M}{100}\ Zn^{2+} \mid Zn$ $E^\circ = 0.1$ volt
e. $Cu \mid M\ Cu^{2+} \parallel \dfrac{M}{100}\ Cu^{2+} \mid Cu$ $E^\circ = 0.1$ volt

17. Given that for the combustion of glucose,
$C_6H_{12}O_6(s) + 6O_2(g) = 6CO_2(g) + 6H_2O(l)$:
$\Delta H = -2820$ kJ
and the combustion of ethanol,
$C_2H_5OH(l) + 3O_2(g) = 2CO_2(g) + 3H_2O(l)$:
$\Delta H = -1380$ kJ
calculate ΔH for the fermentation of glucose,
$C_6H_{12}O_6(s) = 2C_2H_5OH(l) + 2CO_2(g)$.

a. $-$ 5580 kJ
b. $-$ 4200 kJ
c. $-$ 1440 kJ
d. $-$ 60 kJ
e. $+$ 1440 kJ

18. Using the approximate ΔG values in kcal from the table below, which of the follow-

ing methods of reducing the oxides of zinc and titanium would be IMPOSSIBLE?

a. ZnO by Na at 500°C
b. ZnO by C at 1000°C
c. ZnO by H_2 at 1500°C
d. TiO_2 by C at 1800°C
e. TiO_2 by Na at 500°C

ΔG in kJ at	0°C	500°C	1000°C	1500°C
$4Na + O_2 = 2Na_2O$	−732	−670	−460	
$2H_2 + O_2 = 2H_2O$	−503	−418	−335	
$2C + O_2 = 2CO$	−272	−376	−460	−544
$2Zn + O_2 = 2ZnO$	−649	−565	−376	−125
$Ti + O_2 = TiO_2$	−880	−775	−670	−565

19. Which of the following statements do you agree with?

a. When 100 cm³ of conc. HCl is added to 100 cm³ of water the volume becomes 200 cm³.
b. When 200 cm³ of H_2 gas is sparked with 100 cm³ O_2 gas the resultant volume is 300 cm³ (all at the same temperature and pressure).
c. When 100 cm³ of carbon disulphide is added to 100 cm³ of water the volume becomes 200 cm³.
d. When a 3 cm cube of NaCl is added to 100 cm³ of water the volume becomes 127 cm³.
e. When 100 cm³ of conc. H_2SO_4 (density 1·8 g cm⁻³) is added to 100 cm³ of water the density becomes 1·4 g cm⁻³.

20. Which of the following statements about the structure of crystalline sugar do you agree with?

a. A crystalline solid consisting of ions.
b. A giant molecule consisting of atoms held together by covalent bonds.
c. Molecules of $C_{12}H_{22}O_{11}$ held together by a high degree of hydrogen bonding.
d. Molecules of $C_{12}H_{22}O_{11}$ held together by van der Waal's forces.
e. Molecules of $C_{12}H_{22}O_{11}$ held together by covalent bonds.

INORGANIC REVISION 1

Only one answer is correct.

1. An element has these properties:
   i. It is shiny and conducts electricity.
   ii. It forms an acidic oxide.
   iii. Its only chloride is a colourless liquid.
   The element could be:
   a. carbon
   b. silicon
   c. vanadium
   d. magnesium
   e. phosphorus

2. Which is the best method of making a sample of aluminium sulphide in reasonable yield?
   a. Passing hydrogen sulphide into aluminium sulphate solution.
   b. Passing hydrogen sulphide over heated aluminium oxide.
   c. Heating powdered aluminium with sulphur.
   d. Adding ammonium sulphide solution to aluminium chloride solution.
   e. Adding sodium sulphide solution to aluminium potassium sulphate solution.

3. All of these reactions work, but one is not performed industrially to any significant extent. Which is it?
   a. Electrolysing brine.
   b. Burning hydrogen in chlorine.
   c. Reducing zinc oxide with carbon.
   d. Electrolysing fused sodium hydroxide.
   e. Reducing calcium phosphate with carbon.

4. The radius of the potassium atom is 231 pm. The radius of the potassium ion will be:
   a. 13 pm
   b. 133 pm
   c. 231 pm
   d. 234 pm
   e. 258 pm

5. Fluorine is a much better oxidizing agent than iodine. The most probable reason for this is that:
   a. fluorine has the smaller atomic radius.
   b. iodine has the smaller atomic radius.
   c. fluorine is more reactive than iodine.
   d. fluorine is less stable than iodine.
   e. fluoride ions have larger electron attracting powers than do iodide ions.

6. A compound has these properties:
   i. It boils at 76°C.

ii. It does not conduct electricity.
  iii. It is readily hydrolysed by water.
  iv. The solution in water is an electrolyte.
  The compound could be:

  a. NaCl
  b. $MgCl_2$
  c. $AlCl_3$
  d. CCl_4
  e. PCl_3

7. Germanium tetrachloride is most likely to be (at room temperature and pressure):

  a. a white crystalline solid which on heating gives anhydrous $GeCl_4$.
  b. a white crystalline solid which on heating gives germanium dioxide.
  c. a solid which fumes in moist air.
  d. a liquid which fumes in moist air.
  e. a gas which is spontaneously flammable in air.

8. An element forms the following:
  i. An amphoteric oxide.
  ii. A chloride which is hydrolysed by water.
  iii. An alum consisting of white crystals.
  Which element is it most likely to be?

  a. K
  b. Fe
  c. Ga
  d. Ge
  e. Se

9. An element has the electron structure $(1s)^2(2s)^2(2p)^6(3s)^2(3p)^6(3d)^3(4s)^2$. It is:

  a. As
  b. Ca
  c. In
  d. Rb
  e. V

10. An element forms a yellow insoluble sulphide. It also forms complex ions listed as $[X(NH_3)_4]^{2+}$ and $[X(CN)_4]^{2-}$. The element could be:

  a. As
  b. Cu
  c. Cd
  d. Co
  e. Ra

11. An element shows oxidation states of $+2$ and $+4$ in its compounds.
  It forms:
  i. The ion XO_2^{2-}.
  ii. The chloride XCl_2.
  iii. An orange oxide, but most other

compounds are white.
The element is most likely to be:

a. in group $4M$.
b. in group $6M$.
c. an alkali metal.
d. a rare earth.
e. a transition metal.

12. A gas turns damp starch/potassium iodide paper blue.
It MUST be:

a. sulphur dioxide.
b. chlorine.
c. a halogen.
d. a reducing agent.
e. an oxidizing agent.

13. In which reaction is there no change in the oxidation state of nitrogen?

a. Dissolving ammonia gas in water.
b. Thermally decomposing ammonia to nitrogen and hydrogen.
c. Mixing nitric oxide with oxygen.
d. Warming ammonium chloride solution with sodium nitrite solution.
e. Thermally decomposing potassium nitrate.

14. A compound is added to cold dilute sulphuric acid. After filtering, the filtrate is mixed with potassium iodide solution and gives a brown colouration.
The compound which would NOT do this is:

a. H_2O_2
b. BaO_2
c. MgO_2
d. Na_2O_2
e. PbO_2

15. Ethanol warmed with acidified potassium dichromate converts it to aluminium chromium(III) sulphate solution. The alcohol:

a. is a reducing agent.
b. is an oxidizing agent.
c. is a catalyst.
d. removes water and forces the equilibrium of the reaction to the right.
e. floats on the solution and prevents oxidation by the air.

16. As an alternative to ethanol in Question 15 which of these would be most suitable?

a. Chlorine
b. Hydrogen
c. Conc. nitric acid
d. Sulphur dioxide
e. Kerosene

17. Each of these reactions will theoretically liberate bromine from an acidified solution of potassium bromide. One reaction, however, is so slow as to be negligible. Which is it?

a. Bubbling in chlorine.
b. Adding conc. nitric acid.
c. Adding potassium permanganate solution.
d. Adding hydrogen peroxide solution.
e. Bubbling in oxygen.

18. Liquid sulphur dioxide ionises thus:
$2SO_2 \rightleftharpoons SO_3^{2-} + SO^{2+}$
(c.f. $2H_2O \rightleftharpoons OH^- + H_3O^+$).
Which compound will act as a base when dissolved in liquid sulphur dioxide?

a. H_2S
b. $SOCl_2$
c. Na_2SO_3
d. Na_2SO_4
e. $NaHSO_4$

19. Iodine is insoluble in water but dissolves readily in potassium iodide solution. This is caused by:

a. the common ion effect.
b. the formation of KI_2^-.
c. the formation of KI_2^+.
d. the formation of I_3^-.
e. the formation of KI_2.

20. 61·5 g of a saturated solution (at 25°C) of a divalent metal sulphate was made up to 250·0 ml with distilled water. 25·00 ml of this solution produced 1·375 g of dry barium sulphate when treated with excess barium chloride solution.
Calculate the solubility of the metal sulphate at 25°C in g per 100 g of water.
(Ba = 137; S = 32; O = 16)
(Molar mass of the metal sulphate: 148.)

a. 14·2
b. 16·5
c. 17·4
d. 19·2
e. 21·7

INORGANIC REVISION 2

Only one answer is correct.

1. A 2 M solution of a compound in water has a pH greater than 11. The compound could be:
   a. Li_2CO_3
   b. $LiHCO_3$
   c. Rb_2SO_4
   d. $CsHSO_4$
   e. $CsNO_3$

2. Brown manganese dioxide stains are best removed from glassware by placing it in an acidified solution of:
   a. Na_2S
   b. $NaHS$
   c. Na_2SO_4
   d. $NaHSO_4$
   e. Na_2SO_3

3. An 0·1 M solution of a reagent gives a precipitate when carbon dioxide is bubbled in. The reagent could be:
   a. KOH
   b. $CsOH$
   c. $MgSO_4$
   d. $SrCl_2$
   e. $Ba(OH)_2$

4. A reagent when added in excess to acidified potassium permanganate solution at 18°C rapidly gives a colourless solution. The reagent could be:
   a. KI
   b. H_2O_2
   c. $(COOH)_2$
   d. $NaHSO_4$
   e. $Fe_2(SO_4)_3$

5. Which are the best materials to use in order to prepare a pure sample of tin(IV) chloride?
   a. Tin(II) oxide and conc. hydrochloric acid.
   b. Tin(IV) oxide and dilute hydrochloric acid.
   c. Tin and conc. hydrochloric acid.
   d. Tin and dry hydrogen chloride.
   e. Tin and dry chlorine.

6. An element has an electron structure listed as $(1s)^2(2s)^2(2p)^6(3s)^2(3p)^6(3d)^{10}(4s)^2(4p)^6(4d)^2(5s)^2$
   It is:
   a. an element in group $2M$.
   b. an element in group $5M$.
   c. a halogen.
   d. a transition metal.
   e. a rare earth metal.

7. A reagent is added in excess to a mixture of zinc sulphate and aluminium sulphate

176

dissolved in water. It precipitates a compound of only one metal, leaving the other in solution.
The reagent could be:

a. HCl(aq)
b. KOH(aq)
c. NH$_3$(aq)
d. Na$_2$CO$_3$(aq)
e. H$_2$S(g)

8. Which statement is NOT correct?

a. Water oxidizes sodium.
b. Hydrogen oxidizes lithium.
c. Sulphur dioxide oxidizes hydrogen sulphide.
d. Sodium hydroxide oxidizes sodium dichromate.
e. Dilute sulphuric acid oxidizes zinc.

9. When suitably reacted, two reagents produce N$_2$O as shown diagrammatically below.

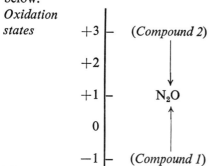

The compounds MUST be:

	Compound 1	Compound 2
a.	HNO$_3$	N$_2$O$_3$
b.	NH$_3$	NO$_2$
c.	N$_2$H$_4$	NO
d.	NH$_2$OH	NaNO$_2$
e.	NH$_4$Cl	NaNO$_3$

10. Analysis of a solid mixture gave these results:

Test	Observation
1. Solubility	Insoluble in cold water and HCl(dil). Soluble in warm excess HNO$_3$(dil).
2. HCl (dil) added to cold solution from (1).	White precipitate forms.
3. Filtrate from (2) added in excess to HgCl$_2$(aq).	White precipitate forms.

The solid mixture could have been:

a. Ag$_2$O + HgO
b. PbO + SnO
c. Bi$_2$O$_3$ + ZnO
d. BaO + MgO
e. CaO + Fe$_2$O$_3$

11. When hydrogen sulphide is passed into an acidic solution, a white turbidity is produced.
The solution could contain:

a. SO_4^{2-}
b. Cl^-
c. Zn^{2+}
d. NO_3^-
e. Mg^{2+}

12. Which reagent does NOT liberate iodine from acidified potassium iodide solution?

a. Sodium nitrite
b. Sulphur dioxide
c. Bromine water
d. Hydrogen peroxide
e. Conc. nitric acid

13. Which aluminium compound is most likely to contain the ion Al^{3+}?

a. Oxide
b. Phosphide
c. Chloride
d. Selenide
e. Fluoride

14. Which of these important processes does NOT involve oxidation or reduction?

a. Charge and discharge of a lead accumulator.
b. Desalination of sea water by ion exchange.
c. Conversion of fuel to electricity in a fuel cell.
d. Corrosion of iron.
e. Discharge of a dry cell battery.

15. An element is burned in oxygen. The products are dissolved in cold dilute hydrochloric acid. Starch/potassium iodide paper dipped into this solution turns blue.
The element is most likely to be:

a. Na
b. C
c. B
d. Al
e. P

16. Chlorine reacts with all the compounds in **a.** to **e.** In which case is the greatest number of moles of chlorine used per mole of reactant?

a. Sodium hydroxide in cold dilute solution.
b. Potassium hydroxide in hot concentrated solution.
c. Aqueous potassium bromide.
d. Ammonia gas.
e. Aqueous iron(II) chloride.

INORGANIC REVISION 2

17. Which compound does NOT exist as a 'giant' structure or ionic lattice?
    a. CO_2
    b. SiO_2
    c. GeO_2
    d. SnO_2
    e. PbO_2

18. A coloured metal sulphate was warmed in solution with excess sodium peroxide. The excess alkali was neutralised with acetic acid. The resulting solution was pale yellow, and addition of lead acetate solution gave a thick yellow precipitate. The metal sulphate could have been:
    a. $CuSO_4$
    b. $NiSO_4$
    c. $MnSO_4$
    d. $Fe_2(SO_4)_3$
    e. $Cr_2(SO_4)_3$

19. All of these elements can be extracted from their compounds by electrolysis. Only one cannot be obtained by any other method. It is:
    a. magnesium
    b. aluminium
    c. bromine
    d. fluorine
    e. zinc

20. 1·32 g of an alloy of aluminium and zinc when dissolved in excess dilute hydrochloric acid liberated 870 cm³ of hydrogen (collected dry at s.t.p.). What was the percentage by weight of aluminium in the sample? (1 mole of gas at s.t.p. occupies 22·4 dm³. Al = 27; Zn = 65.)
    a. 15 per cent
    b. 26 per cent
    c. 35 per cent
    d. 43 per cent
    e. 59 per cent

INORGANIC REVISION 3

Only one answer is correct.

1. An anhydrous powder is heated strongly in a test tube. A solid residue is left. White fumes are produced and damp potassium dichromate paper placed in the tube turns green. The solid could be:
   a. $CuSO_4$
   b. $FeSO_4$
   c. $MgSO_4$
   d. $Al_2(SO_4)_3$
   e. $ZnSO_4$

2. A compound when added to water gives a white turbidity which dissolves when excess concentrated hydrochloric acid is added.
   The compound could be:
   a. $FeCl_3$
   b. PCl_3
   c. NCl_3
   d. $AlCl_3$
   e. $SbCl_3$

3. Two metals are bolted firmly together and placed in dilute hydrochloric acid. In which case would you expect the first named to corrode faster than the second?
   a. Pb and Zn
   b. Fe and Zn
   c. Ag and Zn
   d. Fe and Sn
   e. Fe and Mg

4. Which of these dilute solutions will NOT give a white precipitate when added to dilute barium chloride solution?
   a. Sulphuric acid
   b. Ammonium carbonate
   c. Sodium hydroxide
   d. Sodium oxalate
   e. Lead nitrate

5. An element has these properties:
   i. On heating the hydroxide, a solid oxide forms.
   ii. Ammonium carbonate solution when added in excess to a solution of the chloride gives a white precipitate. This on heating forms the oxide and liberates carbon dioxide.
   The element could be:
   a. Rb
   b. Cs
   c. Pb
   d. Cu
   e. Al

6. Hydrogen sulphide and sulphur dioxide separately decolourise potassium permanganate in acid solution.
   When hydrogen sulphide and sulphur

dioxide are mixed in solution then:

a. there is no reaction as they are both oxidizing agents.
b. there is no reaction as they are both reducing agents.
c. there is no reaction unless they are mixed as gases.
d. the hydrogen sulphide reduces the sulphur dioxide.
e. the sulphur dioxide reduces the hydrogen sulphide.

7. Sulphide stains on silver can be removed by placing the article in contact with a metal in a warm dilute solution. Silver sulphide is then reduced to silver. Which is the best combination of metal and solution?

a. Aluminium and nitric acid.
b. Aluminium and sodium carbonate.
c. Magnesium and sodium hydroxide.
d. Lead and acetic acid.
e. Iron and potassium hydroxide.

8. Which of these important industrial processes does NOT involve reduction or oxidation?

a. Hydrogen from steam and natural gas.
b. Electrolysis of brine.
c. Ammonia-soda (Solvay) process.
d. Extraction of iron.
e. Extraction of bromine from sea water.

9. A drop of water is suspended from a glass rod in the fumes obtained by heating a salt with concentrated sulphuric acid. A gelatinous skin forms on the drop. The salt could be:

a. NaBr
b. CsI
c. LiCl
d. BaF_2
e. $Sr(NO_3)_2$

10. Which reagent would you expect to liberate selenium from hydrogen selenide when this gas is bubbled into an aqueous solution?

a. Cadmium(II) sulphate
b. Lead(II) acetate
c. Magnesium(II) chloride
d. Iron(III) sulphate
e. Copper(II) chloride

11. Copper(II) chloride can be converted to insoluble copper(I) chloride by treating the solution with:

a. sulphur dioxide
b. potassium permanganate
c. chlorine water
d. nitric acid
e. sodium phosphate

12. Which of these reactions would you expect to go to completion in the shortest time when equation quantities are mixed at room temperature?

   a. Zinc sticks are placed in lead acetate solution.
   b. Sulphur dioxide in water reacting with oxygen in the air.
   c. Potassium permanganate and acidified sodium nitrate solutions are mixed.
   d. Bismuth trichloride crystals are hydrolysed by water.
   e. Calcium carbonate powder is added to dilute nitric acid.

13. Which chloride is NOT appreciably hydrolysed by water?

   a. BCl_3
   b. $AlCl_3$
   c. $SnCl_2$
   d. $PbCl_2$
   e. PCl_3

14. Which of these processes does NOT involve the use of a catalyst?

   a. Haber process for ammonia.
   b. Contact process for sulphuric acid.
   c. Lime-soda process for sodium hydroxide.
   d. Production of margarine from an unsaturated oil.
   e. Hydrogen from naphtha and steam.

15. An aqueous solution is added to copper (II) sulphate solution. A precipitate forms which dissolves when excess reagent is added.
    The reagent could be:

   a. NH_4Cl
   b. $(NH_4)_2SO_4$
   c. KCN
   d. KI
   e. NaOH

16. An oxide of an element dissolves in concentrated sulphuric acid giving a yellow solution. Passing in sulphur dioxide turns the solution blue. Further reaction with zinc amalgam turns the solution green and finally pale pink.
    The element is most likely to be:

   a. Ge
   b. Ga
   c. Te
   d. Sb
   e. V

17. Iron containing an impurity is warmed with dilute hydrochloric acid. The gas

produced, when passed through a heated glass tube, gives a silvery deposit near the source of heat.
The impurity could be:

a. Al
b. SnO_2
c. As_2S_3
d. $BeCl_2$
e. PbO

18. A mixture was analysed as follows.

Test	Observation
1 Solubility	Insoluble in water. Dissolves in HCl(dil).
2 Excess NH_3(aq) added to solution from (1).	Coloured solution and white precipitate both formed.
3 Filtrate from (2) has H_2S(g) in.	Black precipitate appears.
4 Precipitate from (2) is dissolved in dil. acid. Excess NH_3(aq) and NH_4Cl(aq) are added followed by sodium phosphate solution.	White crystals slowly form.

a. $CuO + ZnSO_4$
b. $AlCl_3 + MgCO_3$
c. $MnO_2 + Al_2(SO_4)_3$
d. $NiCO_3 + MgO$
e. $CoSO_4 + Na_2CO_3$

The mixture could be:

19. Which statement is correct concerning the soluble salt of formula $K_3Co(CN)_6$? (Atomic number of Co is 27.)

a. It can be named potassium hexacyanocobaltate(II).
b. Co appears not to have a rare gas electron structure.
c. The co-ordination number of Co is 9.
d. Addition of dilute hydrochloric acid gives rapid evolution of HCN(g).
e. The preparation of the compound from $CoCl_2$(aq) and KCN(aq) involves oxidation at some stage.

20. 1·00 g of a metal oxide when completely converted gave 2·33 g of the corresponding sulphate. The mass number of the metal was 91. A formula which corresponds to these results is:
(O = 16; S = 32)

a. M_2O
b. MO
c. MO_2
d. M_2O_3
e. MO_3

ORGANIC REVISION 1

Only one answer is correct.

1. A liquid when warmed with iodine and sodium hydroxide solution gave a yellow precipitate of iodoform (CHI_3). The liquid could have been:
   a. methanol
   b. any primary alcohol
   c. any alcohol
   d. benzyl alcohol
   e. butan-2-one

2. The alkanes constitute a homologous series.
   Which statements are correct?
   a. There is a general increase in boiling and melting points as the series is ascended.
   b. The general formula C_nH_{2n+2} applies to them all.
   c. They have almost identical chemical properties.
   d. All of the above.
   e. None of the above because isomerism is excluded.

3.
$$CH_3-CH_2-\underset{\underset{OH}{|}}{\overset{\overset{H}{|}}{C}}-CH_3$$
   The best name is:
   a. butan-2-ol
   b. butan-3-ol
   c. butanol
   d. tertiary butyl alcohol
   e. methyl propanol

4. C_3H_8O could be:
   a. diethyl ether
   b. propanol
   c. butanol
   d. propanal
   e. butanone

5. A compound was refluxed with excess sodium hydroxide solution. Sodium propanoate was left in solution and ammonia was liberated.
   What could the original compound have been?
   a. C_2H_5CN
   b. CH_3CONH_2
   c. CH_3COONH_4
   d. CH_3CHNOH
   e. $C_2H_5NH_3^+Cl^-$

6. Equal volumes of chlorine and methane are mixed and then left for a long time in daylight.
   What is the product most likely to be?
   a. Chloromethane
   b. 1,2-dichloromethane
   c. Trichloromethane
   d. Carbon tetrachloride
   e. A mixture of all of them with chloromethane predominating.

7. A gaseous hydrocarbon when bubbled into bromine water rapidly decolourised it.
   The hydrocarbon MUST have been:
   a. methane
   b. ethane
   c. but-1-ene
   d. ethylene
   e. any gaseous hydrocarbon containing a carbon-carbon double bond.

8. An attempt was made to prepare ethyl iodide by warming gently "methylated spirits", potassium iodide crystals and concentrated sulphuric acid.
   No ethyl iodide was produced because:
   a. the mixture should not have been warmed.
   b. the alcohol was completely dehydrated to ethylene.
   c. the sulphuric acid should have been diluted.
   d. methyl iodide was produced since methylated spirits is mainly methyl alcohol.
   e. the potassium iodide was oxidized to iodine, which does not react with alcohol under these conditions.

9. If pure 1-iodopropane were to be mixed with excess dilute aqueous sodium hydroxide and refluxed for an hour, we would see in the flask:
   a. propylene gas liberated.
   b. two distinct colourless layers.
   c. two layers with the bottom one coloured purple.
   d. a brown precipitate of iodine.
   e. a homogeneous mixture.

10. A liquid suspected of being an alcohol had a small piece of sodium added. Hydrogen gas was liberated.
    The liquid could have been:
    a. ethanol.
    b. any liquid alcohol.
    c. water.
    d. a mixture of water and a miscible alcohol.
    e. any of the above.

11. A compound C_3H_9N is dissolved in excess dilute hydrochloric acid and sodium nitrite solution is added.
    Bubbles of an invisible odourless gas are given off. The compound could be:
    a. any amine.
    b. $CH_3CH_2CH_2NH_2$
    c. $CH_3CH_2NHCH_3$
    d. $(CH_3)_3N$
    e. All of the above.

12. A solid compound containing nitrogen was boiled with sodium hydroxide solution, leaving a solution of sodium butanoate.
    The compound could have been:
    a. propanamide
    b. 2-aminopropanoic acid
    c. butanamide
    d. butylamine
    e. propylamine

13. One way to analyse an ester is by hydrolysis. The best method is to:
    a. boil with water.
    b. boil with dilute acid.
    c. reflux with dilute acid.
    d. reflux with conc. sulphuric acid.
    e. reflux with dilute sodium hydroxide solution.

14. Methyl cyanide is refluxed with excess sodium hydroxide solution. The resulting solution contains:
    a. acetic acid
    b. ammonium acetate
    c. sodium acetate
    d. acetamide
    e. ethyl isocyanide

15. An alcohol was boiled with acidified sodium dichromate solution and the vapour was passed into warm ammoniacal silver nitrate solution. A black precipitate appeared. The vapour was also passed into a solution of 2,4-dinitrophenylhydrazine, when an orange precipitate appeared.
    The alcohol could have been:
    a. CH_3CH_2OH
    b. $CH_3CHOHCH_3$
    c. $(CH_3)_3COH$
    d. $C_6H_5CHOHCH_3$
    e. C_6H_5OH

16. Compound A is refluxed with excess iodine and red phosphorus to give B.
    B when refluxed with potassium cyanide solution gives C.
    C can be catalytically reduced to D.
    Reaction of D with sodium nitrite and acid gives propan-1-ol.
    A is:
    a. acetic acid
    b. propan-2-ol
    c. propan-1-ol
    d. ethanol
    e. methanol

17. Acetamide can be prepared by the thermal decomposition of ammonium acetate in the presence of glacial acetic acid. The acetic acid:
    a. acts as a catalyst.
    b. lowers the boiling point of the acetamide.
    c. increases the acetate ion concentration, thus producing more acetamide in the equilibrium mixture.
    d. combines with the water formed, thus producing more acetamide in the equilibrium mixture.
    e. inhibits the decomposition of ammonium acetate to ammonia and acetic acid.

18. An organic acid decolourises bromine water. 0·58 g of it needs 100 ml 0·1 M NaOH in a titration. Which acid could it be?

a. $CH_2=CHCOOH$
b. $CH_3CHClCOOH$
c. NH_2CH_2COOH
d. CHCOOH
   ‖
   CHCOOH
e. COOH
   |
   COOH

19. Compound A is refluxed with excess acidified potassium dichromate giving B. B is warmed with excess phosphorus pentachloride giving C. C is shaken with concentrated ammonia solution giving solid D. D is heated with phosphorus pentoxide giving E. E has the molecular formula C_2H_3N. A could be:

a. ethanol
b. propan-1-ol
c. propan-2-ol
d. butanol
e. dimethyl ether

20. 10 cm³ of a gaseous hydrocarbon were sparked in a eudiometer with a large excess of oxygen. There was a decrease in total volume of 20 cm³. On shaking with a solution of sodium hydroxide there was a further diminution in volume of 30 cm³. All volumes were measured at 298 K and 10^5 N m⁻². The gas could have been:

a. C_2H_4
b. C_2H_6
c. C_3H_4
d. C_3H_6
e. C_3H_8

ORGANIC REVISION 2

Only one answer is correct.

1. $CH_3CH(OH)CH_2CH_3$ is best named:
   a. 2-hydroxyisopropyl alcohol
   b. butyl alcohol
   c. *iso*-butyl alcohol
   d. *sec*-butyl alcohol
   e. butan-2-ol

2. Compound A was heated with bromine and excess sodium hydroxide solution. The product B formed a salt with hydrochloric acid and also gave a deep blue colour with copper(II) sulphate solution. A could have been:
   a. ammonium acetate
   b. acetamide
   c. glycine
   d. aniline
   e. acetaldoxime

3. A compound
   i. decolourises bromine water,
   ii. decolourises acidified potassium permanganate,
   iii. gives a precipitate with ammoniacal silver nitrate.
   The compound could be:
   a. propane
   b. but-2-ene
   c. but-1-yne
   d. symmetrical dimethyl acetylene
   e. benzene

4. A compound both effervesces with sodium hydrogen carbonate solution and is miscible with water in all proportions. It could be:
   a. formic acid
   b. carbolic acid
   c. carbonic acid
   d. stearic acid
   e. phenol

5. Which compound is isomeric with propanol?
   a. Acetone
   b. Propanal
   c. Methyl ethyl ketone
   d. Methyl acetate
   e. Methyl ethyl ether

6. Which of the following is miscible with water in all proportions?
   a. Diethyl ether
   b. Dichlorobenzene
   c. Butyl acetate
   d. Propan-2-ol
   e. Benzoic acid

ORGANIC REVISION 2

7. 20 cm³ of a hydrocarbon were sparked with 100 cm³ of oxygen. 80 cm³ of gas remained but after shaking with potassium hydroxide solution 40 cm³ were left. The hydrocarbon could have been:
(All measurements at room temperature and pressure.)

   a. methane
   b. ethane
   c. ethylene
   d. acetylene
   e. propane

8. A compound readily liberates ammonia on warming gently with excess sodium hydroxide solution.
It is most likely to be:

   a. methylamine
   b. ethyl cyanide
   c. acetamide
   d. ammonium oxalate
   e. 1-aminoacetic acid

9. Compound X of molecular formula C_3H_9N forms a white crystalline compound of formula $C_3H_{10}NCl$ with hydrochloric acid. On reaction with nitrous acid, X gives propan-1-ol and nitrogen. What is the structural formula of X?

   a. $CH_3CH_2CH_2NH_2$
   b. $CH_3CH_2NHCH_3$
   c. $(CH_3)_2CHNH_2$
   d. $(CH_3)_3N$
   e. $(CH_3CH_2CH_2NH_3)^+Cl^-$

10. Compound Y of molecular formula C_4H_8O gives a yellow precipitate with 2,4-dinitrophenylhydrazine but does not reduce Fehling's solution.
What could Y be?

    a. C_3H_7CHO
    b. $CH_3CH=C(OH)CH_3$
    c. $CH_3CH_2CH_2CHO$
    d. $(CH_3)_2CHCHO$
    e. $CH_3COC_2H_5$

11. A carboxylic acid forms a methyl ester of molecular formula $C_4H_8O_2$.
The acid is:

    a. formic acid
    b. acetic acid
    c. propanoic acid
    d. oxalic acid
    e. butanoic acid

12. The reduction of an alkyl cyanide produces a compound of formula $CH_3CH_2CH_2NH_2$.
What compound would be produced if the same cyanide were to be hydrolysed with a dilute mineral acid?

    a. Acetamide
    b. Diethylamine
    c. Propylamine
    d. Propanamide
    e. Propanoic acid

13. An ester is to be prepared by refluxing together an organic acid with an alcohol and a catalyst until equilibrium is reached. Which quantities will be certain to give the largest equilibrium yield of the ester?

   a. Equal weights of alcohol and acid.
   b. Equal volumes of alcohol and acid.
   c. 20 ml of acid and 10 ml of alcohol.
   d. Equal numbers of moles of alcohol and acid.
   e. 2 moles of acid and 1 mole of alcohol.

14. Which one of the following does not dissolve appreciably in water, but does dissolve in dilute hydrochloric acid?

   a. Acetamide
   b. Urea
   c. Benzamide
   d. Aniline
   e. 1,3-Dinitrobenzene

15. Which compound will give an intramolecular anhydride?

   a. Acetic acid
   b. Benzoic acid
   c. Phthalic acid
   d. Formic acid
   e. Butanoic acid

16. Acetyl chloride produces acetic acid when hydrolysed. This is best done in the laboratory by:

   a. adding cold water.
   b. heating with water.
   c. boiling with water.
   d. boiling with aqueous sodium hydroxide.
   e. boiling with a dilute acid.

17. A compound was chlorinated in ultraviolet light by passing chlorine into 1 mole of the boiling liquid until there was a weight increase of 69 g. The compound so formed was refluxed with excess alkali and gave another compound of molecular formula C_7H_6O. The original compound could have been: ($Cl = 35.5$.)

   a. toluene
   b. benzene
   c. ethyl benzene
   d. nitrobenzene
   e. phenol

18. The Hofmann bromination (degradation) reaction refers to:

   a. conversion of an amide to an amine.
   b. oxidation with bromine.
   c. making an alkyl bromide.
   d. adding bromine to a double bond.
   e. replacing (OH) by (Br) in a molecule.

19. A compound was passed into chlorine water. The product was boiled with an aqueous alkali giving a compound widely used as an antifreeze.
    The original compound was:

    a. ethane
    b. ethylene
    c. acetylene
    d. propane
    e. propyne

20. Benzene was treated with compound A and aluminium chloride as a catalyst, giving B.
    B was reduced catalytically with hydrogen to C.
    C can be dehydrated to D.
    Polymerisation of D gives the plastic polystyrene.
    A could have been:

    a. chloroethane
    b. iodoethane
    c. ethylene
    d. acetone
    e. acetyl chloride

ORGANIC REVISION 3

Only one answer is correct.

1. Which of these is least likely to be used in the industrial production of organic compounds?
   a. Chlorine
   b. Ammonia
   c. Calcium carbide
   d. Nickel
   e. Potassium permanganate

2. $CH_3CH_2CH=CH_2$ is best named:
   a. 1-butane
   b. but-1-ene
   c. 1-butyne
   d. butylene
   e. 3,4-butyne

3. Which of these acids will probably be the most fully ionised when dissolved in water to make a 0·1 M solution?
   a. Acetic acid
   b. Monobromoacetic acid
   c. Dichloroacetic acid
   d. Trichloroacetic acid
   e. Propanoic acid

4. How many structural isomers are there of the alcohol C_4H_9OH?
   a. 2
   b. 3
   c. 4
   d. 5
   e. 6

5. CH_3CH_2CN can be named:
   a. acetonitrile
   b. ethyl nitrile
   c. cyanopropane
   d. propanonitrile
   e. propyl cyanide

6. Which of these is insoluble in water but dissolves in cold dilute sodium hydroxide solution?
   a. Acetic acid
   b. Stearic acid
   c. Ethyl benzoate
   d. Propylamine
   e. Urea

7. An alcohol when oxidized by acid potassium dichromate produced butanone. The alcohol was:
   a. propan-1-ol
   b. propan-2-ol
   c. butan-1-ol
   d. butan-2-ol
   e. 2-methylpropan-2-ol

8. An organic liquid rapidly decolourised an acid solution of potassium permanganate. This shows that the liquid:
   a. was easily oxidized.
   b. contained a carbon-carbon double bond.
   c. was unsaturated.
   d. was an olefine.
   e. was an alkene.

9. Iodomethane refluxed with magnesium turnings in dry ether gives CH_3MgI. This very useful reaction was first used extensively by:
   a. Hofmann
   b. Fischer
   c. Fittig
   d. Gattermann
   e. Grignard

10. Which compound is NOT an isomer of octane?
    a. 2-methylheptane
    b. 2,3-dimethylhexane
    c. 2,3,4-trimethylpentane
    d. 2,2-dimethylpentane
    e. 2,2,3,3-tetramethylbutane

11. Which substance is NOT usually obtained industrially from ethylene?
    a. Ethylene oxide
    b. Polyvinyl chloride
    c. Ethanol
    d. Glycol
    e. Formaldehyde

12. With which reagent does acetaldehyde react to give a product which is NOT a crystalline solid?
    a. Hydrogen
    b. Ammonia
    c. Hydroxylamine
    d. Sodium bisulphite
    e. 2,4-dinitrophenylhydrazine

13. Compound A is shaken with concentrated ammonia solution giving B. When B is warmed with bromine and excess alkali, a weak base of molecular formula

C_6H_7N is formed.
What could A be?

a. Benzoic acid
b. Toluene
c. Aniline
d. Benzoyl chloride
e. Benzyl chloride

14. A compound was refluxed with excess potassium hydroxide solution for an hour. After acidification with nitric acid, silver nitrate solution was added. No precipitate appeared.
What could the compound have been?

a. CH_3CH_2I
b. $CH_3CH_2CH_2Br$
c. CH_3COCl
d. $C_6H_5CH_2Br$
e. C_6H_5Cl

15. Excess sodium nitrite solution was added slowly to an ice-cold acidic solution of compound A in water. The temperature did not rise higher than 5°C. The mixture was poured into 2-naphthol dissolved in excess alkali, and gave an orange precipitate.
What could A have been?

a. Acetamide
b. Mononitrobenzene
c. Benzamide
d. Aniline
e. Benzylamine

16. Compound A was refluxed with potassium cyanide dissolved in ethanol and gave B. B was reduced to C using hydrogen and a catalyst.
C was dissolved in cold dilute hydrochloric acid.
When sodium nitrite solution was added to this, nitrogen was evolved and the organic product D could be distilled off (in small yield).
D gave propylene when warmed with excess concentrated sulphuric acid.
What was A?

a. 1-bromopropane
b. 2-iodopropane
c. Ethanol
d. Iodoethane
e. Chloromethane

17. Compound A was refluxed with excess sodium hydroxide solution giving sodium iodide and B.
B was oxidized with concentrated acidified

potassium dichromate solution to C. C (which was insoluble in water) was dissolved in excess sodium hydroxide solution and when evaporated to dryness left solid D.
D was heated strongly with soda-lime and gave benzene (in poor yield).
What was A?

a. C_6H_5I
b. $C_6H_5CH_2I$
c. $IC_6H_4CH_3$
d. $C_6H_5C_2H_5$
e. IC_6H_4I

18. A compound was oxidized with concentrated acidified potassium dichromate solution. The organic acid produced, after separating and drying, was heated. 1 mole of it lost 1 mole of water.
The original compound could have been:

a. 1,2-dimethylbenzene (ortho-xylene)
b. 1,3-dimethylbenzene (meta-xylene)
c. 1,4-dimethylbenzene (para-xylene)
d. methylbenzene (toluene)
e. ethylbenzene

19. 1 mole of compound A was mixed with 1 mole of hydrogen iodide gas, giving B. B was refluxed with excess sodium hydroxide solution to give C.
C was oxidized with acidified potassium dichromate solution to give D.
D had the molecular formula C_3H_6O, and gave no reaction with Tollens' or Fehling's reagents.
What could A have been?

a. $CH_2=CH_2$
b. $CH_3CH=CH_2$
c. $CH_3CH_2CH=CH_2$
d. $CH_2=CH—CH=CH_2$
e. $H—C\equiv C—H$

20. When completely oxidized, 0·26 g of a compound gave 0·88 g of carbon dioxide and 0·18 g of water. The compound gave no reaction with bromine water or ammoniacal silver nitrate solution.
It could have been:
($H = 1; C = 12; O = 16$.)

a. ethane
b. ethylene
c. acetylene
d. benzene
e. styrene

ANSWERS AND COMMENTS

ACID – ALKALI TITRATIONS

Answers and comments

The concepts of equivalents and normalities have rapidly lost favour but there is no doubt that practising chemists have found them most convenient and many will continue to use the terms for some time to come. Throughout any transition period to moles and molar solutions it is essential that there should be facility in quick conversions.

Simply:

$$1 \text{ g-equivalent of an acid} = \frac{1 \text{ mol of the acid}}{n}$$

$$1 \text{ litre of N acid} = \frac{1}{n} \text{ litres of M acid}$$

where n is the basicity of the acid.

1. a, c, e
2. d, e
3. a, c
4. a, c, d
5. b
6. a, b, d
7. c, e
8. b, d
9. a
10. b
11. e
12. d
13. d
14. c

Since 0·216 g of silver is equivalent to 0·362 g of silver salt, 108 g of silver are equivalent to 181 g of silver salt. The acid has an equivalent of 181 – 107 or 74.
The molar mass of the acid is below 100 so that it cannot be other than monobasic. The molar mass of the ethyl ester of propanoic acid is 102.

15. a, c

1·7 moles of NH_3 occupy 38·08 litres at s.t.p. if 1 mole occupies 22·4 litres.

16. c, d

250 ml of solution contained $\frac{20}{1000}$ mol of NaOH.
Since the mol is 40 g the solution contained 0·8 g of NaOH.

ANSWERS AND COMMENTS

17. e

The solution contained $\dfrac{37\cdot 5}{1000}$ mol of alkali in 250 ml.

If there are $\dfrac{x}{1000}$ mol of KOH and $\left(37\cdot 5 - \dfrac{x}{1000}\right)$ mol of NaOH, then there will be $\dfrac{56x}{1000}$ g of KOH and $\dfrac{40\,(37\cdot 5 - x)}{1000}$ g of NaOH

$$\dfrac{56x}{1000} + \dfrac{40(37\cdot 5 - x)}{1000} = 1\cdot 7$$

$16x = 1700 - 1500$

$x = 12\cdot 5$

i.e. $12\cdot 5 \times 10^{-3}$ mol of KOH
$25\cdot 0 \times 10^{-3}$ mol of NaOH

18. a

The difference between the two titrations is 8 ml 0·1 M HCl which is required to convert $NaHCO_3$ to NaCl.
i.e. 16 ml 0·1 M HCl to convert the Na_2CO_3 to NaCl, and 8 ml 0·1 M HCl to convert the NaOH to NaCl.

The solution of alkali is thus $\dfrac{8}{250}$ M with respect to Na_2CO_3 and $\dfrac{8}{250}$ M with respect to NaOH.

19. b, c, d

$\dfrac{15}{1000}$ mol of NH_3 are produced so that the original contained $\dfrac{15}{1000} \times 14$ g of nitrogen.

20. c, d

16 ml of M acetic acid are produced by the hydrolysis of the ester which is $\dfrac{16}{1000}$ mol, i.e. 16 per cent of the original number of moles of ester.

OXIDATION NUMBERS AND REDOX TITRATIONS

Answers and comments

The oxidation of an element, alone or in a compound, can be considered as an increase in its oxidation state or alternatively as the loss of one or more electrons from an atom, ion, or molecule.

e.g. $MnO_2 \rightarrow MnO_4^-$

a. Oxidation state of Mn in MnO_2 is $+4$
Oxidation state of Mn in MnO_4^- is $+7$
Increase in the oxidation state is $+3$

b. $MnO_2 + 4OH^- \rightarrow MnO_4^- + 2H_2O + 3e^-$
Three electrons are transferred to the oxidizing agent.

1. d
2. d
3. e
4. a, b, c, d, e
Oxidation occurs at the anode; i.e. the anode always dissolves both in electrolytic and in primary cells. Hence the phrase 'anodic corrosion'.

5. d
6. b
7. b, e
The co-ordination number of molybdenum is 8 since there are 8 co-ionic bonds between molybdenum and the cyanide ions.

8. c
Although **a** is also an oxidation, if the nitrite is added to the acid permanganate the NO_3^- ion is produced. This reaction is used in the estimation of nitrite, the nitrite solution always being added to the permanganate solution.

9. a, c
Using ionic half equations
$$Cr_2O_7^{2-} + 14H^+ + 6e^- \rightarrow 2Cr^{3+} + 7H_2O$$
$$6Fe^{2+} \rightarrow 6Fe^{3+} + 6e^-$$

10. a, d
0·56 g of Fe^{2+} is 0·01 mol and will reduce $\frac{0·01}{6}$ mol of dichromate or $\frac{0·01}{5}$ mol of permanganate ions.

ANSWERS AND COMMENTS

11. d, e

Since hydrogen peroxide is oxidized, electrons must be transferred from the peroxide to the permanganate. **d** and **e** amount to the same thing when the half equations are added for peroxide and acid permanganate.

12. a, b

Most oxidizing agents convert thiosulphate to sulphate, iodine being unusual in converting it to tetrathionate.

13. b, c

For the titration the ionic half equation is
$$H_2O_2 \rightarrow 2H^+ + O_2 + 2e^-$$
The catalytic decomposition of hydrogen peroxide (by which the volume strength is measured) proceeds according to the equation
$$2H_2O_2 \rightarrow 2H_2O + O_2$$

14. e

This is the only case in which there is a change in oxidation state.

15. b, e

In the reaction of potassium permanganate with oxalic acid we have the two half equations:
$$C_2O_4^{2-} \rightarrow 2CO_2 + 2e^-$$
$$8H^+ + MnO_4^- + 5e^- \rightarrow Mn^{2+} + 4H_2O$$
so that the 1 mole of oxalic acid is oxidized by 0·4 mol $KMnO_4$, and 25 ml 0·1 M oxalic acid are oxidized by 10 ml 0·1 M $KMnO_4$.
In the reactions of oxalic acid with alkali, since oxalic acid is dibasic it will react with 1 mole of barium hydroxide but with 2 moles of sodium hydroxide.

16. d, e

$$C_2O_4^{2-} \rightarrow 2CO_2 + 2e^-$$
$$Fe^{2+} \rightarrow Fe^{3+} + e^-$$

From these equations it is seen that 1 mol FeC_2O_4 gives up 3 mol of electrons on oxidation, so that 5 mol FeC_2O_4 reduces 3 mol $KMnO_4$

17. a, b, c

201

ANSWERS AND COMMENTS

18. b, c, d, e

$$2S_2O_3^{2-} \rightarrow S_4O_6^{2-} + 2e^-$$
$$I_2 + 2e^- \rightarrow 2I^-$$

So that 2 mol $Na_2S_2O_3$ reduces 1 mol I_2 and 100 ml 0·1 M $Na_2S_2O_3$ reduces 50 ml 0·1 M I_2.

19. a, c, e

20. a, d

25 ml 0·1 M solution of reduced vanadium are oxidized by 15 ml 0·1 M potassium permanganate, *so* 1 litre of the reduced vanadium solution is oxidized by 3/5 litre of potassium permanganate, *and* 1 mol of reduced vanadium is oxidized by 3/5 mol of potassium permanganate. Since the oxidation state of manganese is reduced by five [Mn^{VII} to Mn^{II}] then the oxidation state of vanadium must be increased by three. Since the final state is +5 the reduced state must have been +2.

MOLECULES, MOLES AND EQUILIBRIA

Answers and comments

1. e
2. a, b, c
  Mass of oil $= 0\cdot 8$ g
  Number of moles $= \dfrac{0\cdot 8}{480} = \dfrac{1}{600}$
  Number of molecules $= \dfrac{6 \times 10^{23}}{600} = 10^{21}$
  These 10^{21} molecules occupy 1 cm³.
  There are 10^7 molecules in 1 cm and the thickness of the film is 10^{-9} cm. Therefore the area of the film is 10^7 cm² or 10^3 m².

3. a, b, d
4. a
5. a
6. b
7. d
8. b, c, d
9. a, b, c, d, e
10. c
  The apparent molar mass is 140
  therefore $\dfrac{1+\alpha}{1} = \dfrac{210}{140}$ where α is the degree of dissociation
  $\alpha = \dfrac{70}{140} = \dfrac{1}{2}$

11. b, e
  Although it seems obvious that all the molecules of acetic acid have dimerised, it must not be assumed that this is the only possibility.

12. a, c, d
  It is often more convenient to consider the partial pressures of gases rather than concentrations since K_p, the equilibrium constant at constant pressure, is more often used.
  Partial pressure of $H_2 = \dfrac{\text{moles of } H_2}{\text{total moles}} \times \text{total pressure}$
  Here there is 0·1 mole of H_2
  0·05 mole of N_2
  0·1 mole of O_2
  0·1 mole of He

ANSWERS AND COMMENTS

     0·05 mole of Cl_2
i.e. 0·4 mole in all.
Since $2·5 \times 10^4$ Nm^{-2} is ¼ of the total pressure, 0·1 mole of gas must give a partial pressure of $2·5 \times 10^4$ Nm^{-2}.

13. a, b, c, e
Although the Equilibrium Law was first applied only to homogeneous systems it has since been applied to heterogeneous systems, ionic systems, and purely physical systems.

14. d, e

15. a, c, d, e

16. b, d

17. b, c
The answers to the last three questions can all readily be seen by the use of Le Chatelier's Principle.

18. a, b, c
Reactions which are endothermic should give a better yield of the product in the equilibrium mixture at a higher temperature. The reaction also goes faster at the higher temperature so that the production of NO should not require a catalyst. We have assumed that the value of ΔH does not change at a high temperature, but this may be a very wrong assumption since the values quoted are for 298 K.

19. b, c

20. a
Much information can often be gained by an examination of the number of molecules on each side of the equation.
e.g. in (iv)
$$K_c \propto \frac{1/v^2}{1/v^4} \propto v^2$$
$$K_p \propto \frac{p^2}{p^4} \propto \frac{1}{p^2}$$

COLLIGATIVE PROPERTIES

Answers and comments

1. a, b, c
2. c
3. b, c, d, e
4. c

Mole fraction of solute $= \dfrac{0\cdot 2}{1}$

Mole fraction of solvent $= \dfrac{0\cdot 8}{1}$

Vapour pressure of solution $= 0\cdot 8 \times 2\cdot 60$ kN m^{-2}
$= 2\cdot 08$ kN m^{-2}

5. b

Moles of solute $= \dfrac{62}{62} = 1$

Moles of solvent $= \dfrac{90}{18} = 5$

Vapour pressure of solution $= \dfrac{5}{6} \times 2\cdot 4$ kN m^{-2}
$= 2\cdot 0$ kN m^{-2}

6. d

Moles of sucrose $= \dfrac{100}{342}$

Moles of glycerol $= \dfrac{60}{92}$

Moles of glucose $= \dfrac{60}{180}$

Moles of glycol $= \dfrac{50}{62}$

Moles of fructose $= 0\cdot 5$

Examination shows that the molar concentration of glycol is the highest:
i.e. the lowest weight of the additives listed produces the greatest lowering of the freezing point.

7. b, c, d

1 g of urea in 30 g of water is the same as 60 g of urea in 1800 g of water or 1 mole of urea in 100 moles of water.

Therefore the mole fraction of solvent is $\dfrac{100}{101} = 0\cdot 99$

ANSWERS AND COMMENTS

1 mole in 22·4 litres of solution gives an osmotic pressure of 1 atmosphere. 1 mole in 1·8 litres of solution gives an osmotic pressure of $\frac{22\cdot4}{1\cdot8}$ atmospheres i.e. about 12 atmospheres.

8. b

There is $\frac{1}{40}$ mole in 100 g or ¼ mole in 1000 g of each solvent.

9. c, d

Ionic dissociation takes place in the salts but can be neglected in the case of acetic acid. Acetyl chloride reacts with water to give acetic acid.

10. a, c, e

−0·96°C is four times as big as −0·24°C so that the solution must be four times as concentrated at this temperature and 75 g of ice are formed.

11. b, d

3 g of pentose is $\frac{1}{50}$ mole in 100 g of water. This is about 1 mole in 5 litres of solution so the osmotic pressure is about 4·5 atmospheres.

12. a

The benzene vapour will have a temperature of 80·4°C. As benzene boils off, the solution becomes richer in benzoic acid and the boiling point of the remaining liquid will rise.

13. c

14. b, d, e

There must have been 1 mole in $\frac{22\cdot4}{14}$ or 1·6 litre of solution. This is about 0·6 mole in 1 litre of solution or 1000 g of water.

15. d

$\frac{6\cdot0}{60}$ mole of urea + $\frac{9\cdot2}{92}$ mole of glycerol = 0·2 mole.

16. a, c, e

The errors must be less than 3 per cent of the measured values. 5 g of the compound in 100 g of solvent is about (i) $\frac{1}{60}$ mole of solute in 100 g of solvent, (ii) 4 moles of solute in

ANSWERS AND COMMENTS

22·4 litres of solution, (iii) mole fraction of solute in aqueous solution is 0·003.
Hence the **a.** freezing point depression of water is about 0·3°C and 3 per cent of 0·3°C is 0·01°C.
b. boiling point elevation is about 0·08°C and 3 per cent of 0·08°C < 0·01°C (read on a Beckmann thermometer).
c. osmotic pressure is about 3000 mmHg and 3 per cent of 3000 mmHg > 10 mmHg.
d. vapour pressure lowering is 0·04 mmHg and 3 per cent of 0·04 mmHg < 0·1 mmHg.
e. freezing point depression of camphor is 7°C and 3 per cent of 7°C > 0·01°C.

17. d, e

$\frac{2}{148}$ mole in 100 g of acetic acid should lower the freezing point by $\frac{39}{74}$°C. Since the experimental value is smaller than this there are fewer particles than expected and association must have taken place. If there is 52 per cent conversion into double molecules the number of effective particles is only $26 + 48$ per cent. This will give a freezing point depression of $\frac{39 \times 74}{74 \times 100}$ or 0·39°C. Similarly it can be shown that **e** is also a possibility.

18. d

$\frac{7 \cdot 2}{366}$ mole of CdI_2 should lower the freezing point of 100 g of water by 0·36°C. This agrees with the experimental figure so that either CdI_2 is undissociated into ions or it exists in some form such as **d**. The latter would appear to be more likely since we would expect such a salt to be fully ionized in aqueous solution.

19. b

From the experimental results the molar mass appears to be 102 ± 10. However, the formula masses appear to be 379, 420, 403, 386, and 486 respectively. On counting the number of ions formed in each possibility the apparent molar masses would be $379, \frac{420}{4}, \frac{403}{3}, \frac{386}{2},$ and $\frac{486}{3}$ so that **b** most nearly agrees.

20. a

The choice is really between $\frac{1}{60}$ mole of urea, $\frac{1}{58 \cdot 5}$ mole of NaCl or $\frac{1}{111}$ mole of $CaCl_2$ but because of ionic dissociation the depressions will appear due to $\frac{1}{60}$ mole of urea, $\frac{2}{58 \cdot 5}$ mole of NaCl, $\frac{3}{111}$ mole of $CaCl_2$. The largest of these is **a**.

ACIDS AND ACIDITY

Answers and comments

1. a, b, c, d

2. a, b, d, e
The fundamental work of Arrhenius shows that for a weak electrolyte the degree of ionization increases as the solution is diluted and only ceases at infinite dilution. A strong electrolyte appears to do much the same, at least when the solution is first diluted. With a weak electrolyte, such as acetic acid, there is a continuous formation of ions from molecules on dilution. A strong electrolyte such as hydrochloric acid is considered to be 100 per cent ionized in aqueous solution but does not behave as such in relatively concentrated solutions because of inter-ionic attractions.

3. a

4. a, b, d
One mole of calcium chloride produces one mole of Ca^{2+} and 6×10^{23} Ca^{2+} ions. Calcium chloride is fully ionized.

5. b, d
Most weak acids are less than 5 per cent ionized at this dilution, whereas strong acids are considered to be 100 per cent ionized. Without precise calculation it is seen that the H^+ concentration in M/100 HCl will be much greater than that in M/10 acetic acid.

6. d
The other quantities depend on concentration.

7. e
The other compounds are all strong electrolytes.

8. b, c, d
In aqueous solution at 25°C, $[H^+][OH^-] = 10^{-14}$ and is called K_w. This, using the notation $pK_w = -\log K_w$, becomes $pH + pOH = pK_w = 14$.

9. a, b, c, d
Using Le Chatelier's Principle we would expect the position of equilibrium to be displaced so that more molecules of water would be formed at the lower temperature, hence K_w decreases as do $[H^+]$ and $[OH^-]$. Consequently the pH rises, but this does not mean that neutral water becomes alkaline as the temperature drops, since the pOH rises by an equivalent amount. The often quoted value of the pH of water as 7·0 only applies at 25°C.

10. a, c, d

ANSWERS AND COMMENTS

11. a, b, c, d

12. a, b, d

These last three questions give practice in the use of the dissociation constant of an acid. In Question 12, pK_a of formic acid is 3·8 or $K_a = 10^{-3 \cdot 8}$
but $10^{-3 \cdot 8} = 10^{-4} \times 10^{0 \cdot 2}$
$\therefore K_a = 1 \cdot 58 \times 10^{-4}$

For a rough approximation $\dfrac{[H^+][\text{formate ion}]}{[\text{undissociated acid}]} = \dfrac{[H^+]^2}{\text{molarity of the acid}}$

$pK_a = 2pH - 3$
$\therefore pH = 3 \cdot 4$

This method may be very satisfactory for the weaker acids like acetic acid and propionic acid but is not suitable for formic acid, because α is not small. If we take the value for the pH that we have just calculated as 3·4 and apply it in the expression $[H^+] = \alpha/V$
then $\alpha = 10^{-0 \cdot 4}$

Hence, in this case we should use Ostwald's Dilution Law in the form

$K_a = \dfrac{\alpha^2}{(1-\alpha)V}$ (where V is the dilution in litres).

13. a, b, c, d

14. e

15. b, e

16. a, b, c

17. a

$K_a = \dfrac{[H^+][\text{acetate ions}]}{[\text{undissociated acid}]}$ We can assume that all the acetate ions came from the sodium acetate.

$[H^+] = K_a \times \dfrac{\text{moles of acetic acid}}{\text{moles of sodium acetate}}$

$= 10^{-4 \cdot 8} \times 10$

$pH \simeq 4$

18. d

Acid salts are not necessarily acidic, neither are normal salts neutral in solution. Hydrolysis will occur whenever we have a salt of a weak acid or a weak base.

19. c

The indicator can be considered to be a weak base in equilibrium with its ions
i.e. $MOH \rightleftharpoons M^+ + OH^-$

Hence $K_b = \dfrac{[M^+][OH^-]}{[MOH]}$

The neutral colour would be expected to occur when $[M^+] = [MOH]$, i.e. when $K_b = [OH^-]$; hence the pH at the neutral colour is 9.
In practice most indicators change over a range of 2 pH units.
One would expect the change here to occur between 8 and 10.

20. a, b, d, e
Such approximate calculations enable us to plot curves for the titration of acid against alkali.

IONIC EQUILIBRIUM – SOLUBILITY PRODUCT – STABILITY CONSTANT

Answers and comments

1. c

The best statement is the one which emphasises the equilibrium which exists between the solid NaCl and the hydrated Na^+ and Cl^- ions.

2. c, d

3. c

These last two questions emphasise that the solubility product principle requires an equilibrium between the solid and the ions in solution. All the other answers are inapplicable.

4. d

Here again, the equilibrium is between the solid barium sulphate and the aqueous ions.

5. d

6. a, b, c, d, e

In 10 litres there is 0·1 mole of NaCl, 0·5 mole of NaBr and 0·4 mole of $BaCl_2$ so that $[Br^-] = 0·5$ mol l^{-1} $[Na^+] = 0·5 + 0·1 = 0·6$ mol l^{-1}, and $[Cl^-] = 0·1 + 0·8 = 0·9$ mol l^{-1}.

7. a, d, e

The drop is diluted $1000 \times \dfrac{1}{0·05} = 20\,000$ times.

$[AgNO_3] = M/20\,000$ or 5×10^{-5} mol l^{-1}.

If $[Ag^+] = 5 \times 10^{-5}$ mol l^{-1} and $[Cl^-] = 1$ mol l^{-1}, then the product of these ion concentrations vastly exceeds the solubility product 10^{-10}.

8. a, b, c, d

Carbon tetrachloride is the only compound which does not provide chloride ions directly or on rapid hydrolysis.

9. a, c, d

10. c, d

The solubility product of $RaSO_4$ is $(6·5 \times 10^{-8})^2$ or $4·2 \times 10^{-15}$.
The solubility of $BaSO_4$ is 1×10^{-5} mol l^{-1}.
Without further information one would expect both salts to be more soluble at 50°C than at 25°C and hence the solubility product should be greater.

11. a, c

$[F^-]$ is 4×10^{-4} mol l^{-1} so that the solubility product becomes $3·2 \times 10^{-11}$.

12. a, c, e

ANSWERS AND COMMENTS

13. a

14. b
Silver bromide is the most insoluble of the first three, which are binary electrolytes. (Solubility about 10^{-6} mol l^{-1}.)
A rough approximation of the solubility of the other two gives
$[Ag_2CrO_4] = 10^{-4}$ and $[Ag_3PO_4] = 10^{-5}$ mol l^{-1}.

15. e

16. a

17. d

18. b, e
$[H^+]^2[S^{2-}] = 10^{-6} \times 10^{-17}$
For precipitation to take place $[Cd^{2+}]$ must be greater than 5×10^{-8} mol l^{-1} and $[Mn^{2+}]$ must be greater than 10^2 mol l^{-1}. Hence only CdS is precipitated and $[Cd^{2+}]$ left in solution will be about 10^{-8} mol l^{-1}.

19. c, d, e
In the first three answers the ionic product is 10^{-13}, and in the fourth and fifth, 10^{-19}. The ionic product exceeds the solubility product in the case of the zinc, chromium(III) and iron(III) hydroxides. It is interesting to note that if these quantities are correct, then we would expect zinc hydroxide to precipitate but the zinc ion forms a complex ammine with ammonia.

20. a, b, c, d
Although the fact that ions are hydrated has been ignored, this would have little effect on the calculations.

ELECTROCHEMISTRY

Answers and comments

1. a, b, d
2. a
3. a
4. e
5. a, b, d
6. c, d

The anode is the electrode at which electrons leave a cell no matter whether it is an electrolytic or a primary cell. It is the electrode which corrodes and which is oxidized (or at which oxidation takes place).

$Zn \rightarrow Zn^{2+} + 2e^-$ in the Daniell cell,
$4OH^- \rightarrow 2H_2O + O_2 + 4e^-$ in the electrolysis of water.

7. a

Examination of the electrode potentials shows that 0·77 volt from the Fe^{3+}/Fe^{2+} equilibrium is large enough to reverse the $I_2/2I^-$ reaction but not the $Br_2/2Br^-$ reaction.

8. a, b, c, d, e
9. b, c, d

Examination of the redox potentials shows that a very high potential is required to liberate sodium compared with hydrogen. Consequently hydrogen is liberated although the hydrogen ion concentration is very small. This disturbs the ionic equilibrium of the water in the vicinity of the cathode and because hydrogen ions are discharged as hydrogen gas, more hydroxide ions are liberated in the vicinity of the cathode.

10. d

Iodine is liberated at the anode:
$2I^- \rightarrow I_2 + 2e$

11. b, e

To increase the e.m.f. of the cell we must increase the tendency for the forward reactions to occur:
i.e. $Zn \rightarrow Zn^{2+} + 2e^-$
and $Cu^{2+} + 2e^- \rightarrow Cu$
Application of Le Chatelier's Principle will show that the forward tendencies are increased by making $[Zn^{2+}]$ as small as possible and $[Cu^{2+}]$ as large as possible.

12. a, b, c

Always making the left-hand electrode the negative, then in **a**
$$E = E_{Mg} + E^{\ominus}_{Zn}$$
$$= 2·37 - 0·76 = 1·61 \text{ volts}$$

ANSWERS AND COMMENTS

13. a, e

A cell is formed in which zinc is the electrode which dissolves forming Zn^{2+} ions. It is thus oxidized and forms the anode (anodic corrosion).

14. b, d, e

15. a, b, c, d

These last two questions illustrate the relationship between the half-cell potential and the concentration of the electrolyte.

16. a, c, d, e

17. b, c, d

About $-2\cdot 1$ volts are required to form sodium amalgam. The hydrogen overvoltage at a mercury electrode is high (about $-1\cdot 3$ volts) and the pH around the cathode is high (about 11 or 12) so that the total value of E which has to be overcome in order to liberate hydrogen is about the same as that for producing the sodium amalgam. Some hydrogen is liberated at the same time as the amalgam is formed.

18. a, b, d, e
19. b, c
20. a, b, c

ENERGETICS

Answers and comments

1. c, d
2. a, c, d
3. a, b, c, d, e
4. a, c
5. a, b, c

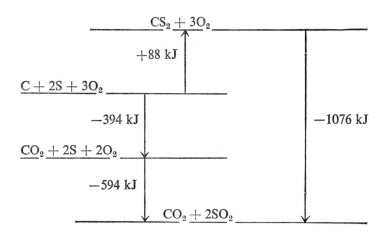

6. a, b, c, d

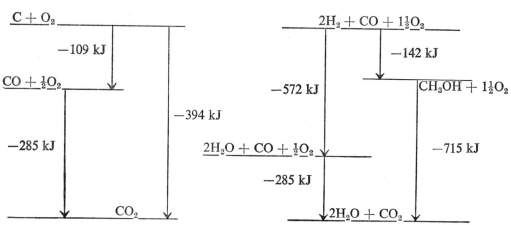

ANSWERS AND COMMENTS

7. a, b, e

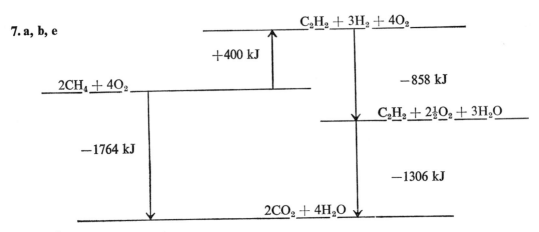

8. b, c, d, e

9. a, b, c, d
Although Le Chatelier's Principle suggests that the reaction should take place at a low temperature, most reactions need a certain activation energy before they can take place.

10. a, b, c, d, e
Energy is lost from the system when bonds are made, i.e. ΔH is negative.
Energy is gained by the system when bonds are broken, i.e. ΔH is positive.

11. a, b, c, e
12. a, b, c
13. a, b, c, e
14. a, b, c, d

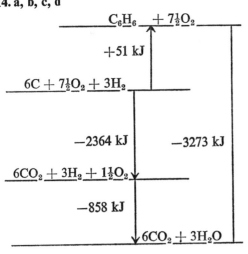
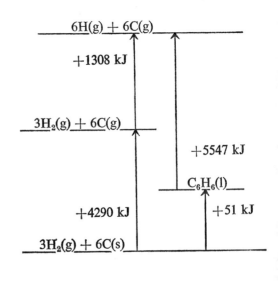

ANSWERS AND COMMENTS

15. a, c, d, e
16. a, b, c, d, e

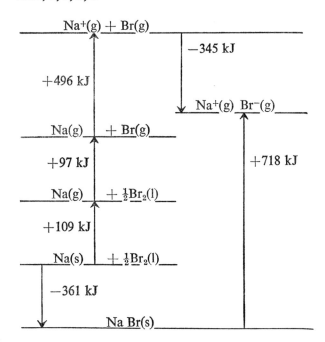

17. e

The quantities given are approximately 0·1 mole of Cu to 0·2 mole of Ag^+. Any increase in the amount of copper would not result in any further reaction without a corresponding increase in the Ag^+ ions.

18. b, c, d

When 0·1 mole of copper dissolves, 0·2 mole of silver precipitates and 0·2 faraday is transferred.

The electrical energy = 0·2 × 96 500 × 0·45 joules for $\frac{1}{10}$ mole of Cu
    i.e.   86 850 joules/mole of Cu

The free energy $(\Delta G) = 86·85$ kJ mol^{-1}

19. a, c, d, e

20. a

Just as 'up-down' diagrams help us to determine ΔH_f so we can use diagrams to find ΔG_f. In those cases in which ΔG_f is negative we can say that the reaction is possible, although we cannot say whether it will take place at a reasonable speed or not.

PHASE EQUILIBRIA

Answers and comments

1. a, b, c

2. a, b, d

3. a, c

4. a, b, c

The metals which would seem most likely to form solid solutions would be transition metals with similar atomic radii. Elements such as tin and lead, in the same group of the periodic classification, would be more likely to form simple eutectic mixtures.

5. c

If all the liquid is allowed to solidify, then obviously the composition of the solid is the same as that of the liquid. The composition of the solid which separates at any given temperature is determined by drawing horizontal 'tie lines' from the upper liquidus curve to the lower solidus curve so that the composition of the solid which first separates is R: i.e. about 90 per cent nickel.

6. b, c, d, e

The mixture of 2 parts of lead to 1 part of tin is a typical plumber's solder for wiping joints in lead pipe whereas the best tinman's solder which sets quickly is 2 parts tin to 1 part lead (i.e. the eutectic mixture).

7. c, d

It should not seem surprising that two metals such as magnesium and tin should form a compound of formula Mg_2Sn since magnesium is a group II metal and tin is in group IV.

8. b, c, e

Here we should not be surprised that a compound is formed since phenol is a weak monobasic acid and 4-aminotoluene is a monacid base.

9. a, c

The vapour over the boiling liquid is richer in the more volatile component. The liquid of composition P boils at T_3 on the lower liquidus curve and the composition of the vapour is found by drawing a horizontal tie line to the upper vapour curve. The liquid remaining in the flask now becomes steadily richer in the less volatile component and the temperature of the boiling mixture rises steadily until it finally becomes T_2. If we wish to separate a mixture then we must take fractions boiling between chosen temperature limits.

10. c

The formation of constant boiling mixtures, or azeotropes, is a very common phenomenon with liquids which might be expected to show other evidence of hydrogen bonding. All too often this is forgotten and distillation is quoted as a method of separation when an examination of the curves shows that it is impossible to obtain both pure components.

ANSWERS AND COMMENTS

11. c, d

12. b, d

Both nitric acid and hydrochloric acid form maximum boiling point mixtures with water and the constant boiling mixture of hydrochloric acid makes an excellent standard acid of known concentration for volumetric analysis.

13. a, b, c, d

There are two compounds of water and nitric acid which can be crystallised.

14. a, b, c

Here we are dealing with two immiscible liquids which boil when the sum of their vapour pressures is the same as the atmospheric pressure, hence the mixture boils at a temperature lower than the boiling point of either. The method of steam distillation is of great value in the distillation of essential oils.

15. c

It is often convenient to be able to make quick and reliable estimates of the vapour pressures of liquids at different temperatures.

16. a, b

If the pressure of water vapour in the atmosphere exceeds the vapour pressure of a saturated aqueous solution of a salt, then crystals of the salt will deliquesce. If the pressure of the water vapour in the atmosphere is less than the vapour pressure of a *lower* hydrate then the salt will effloresce. Since $MgSO_4,7H_2O$ has a vapour pressure of 11·5 mmHg and $MgSO_4,6H_2O$ has a vapour pressure of 9 mmHg, water will be lost by the $MgSO_4,7H_2O$.

17. a, e

18. a, b, c, d

19. b, e

The experimental results are in agreement with $\frac{\text{(concentration in water)}^2}{\text{concentration in CCl}_4}$ being constant. This could be interpreted as the acid being 100 per cent dissociated into ions in the water, or the acid associating in the CCl_4 phase. The former is completely unreal.

20. a, b, c, d

The Distribution Law provides a useful method of determining the formula of a complex ion. The titrations give the information that there is $4·5 \times 10^{-2}$ mole of NH_3 in 50 ml of aqueous solution but that the *free* NH_3 in the aqueous layer is twenty-five times the NH_3 in the CCl_4 layer. The difference is the NH_3 which is associated with the Cu^{2+} ions. There are 4 moles of NH_3 to every mole of Cu^{2+} ions.

STATES OF MATTER

Answers and comments

1. d
2. b, c, d, e
3. a, b, c, d, e
4. a, e
5. c, d, e
   Using Graham's Law the vapour density of the gas will be 49 times that of hydrogen, i.e. the molar mass will be about 98. Since the molar mass of $COCl_2$ is 99 this seems a very likely possibility.
   The gas will diffuse at $\left(\dfrac{16}{49}\right)^{\frac{1}{2}}$ or $\dfrac{4}{7}$ the speed of oxygen.

6. a, c, e
7. a, b, c, d, e
8. a, b, c
9. a, b, c
   The limiting radius of the smaller ion is determined on the assumption that when the large ions come into contact the structure is likely to change to the one of next lower co-ordination number. It is better if the terms 'body and face centred' are confined to lattices of atoms with the same radii.

10. a, b, e
11. e
    The false step is in the last line. We have equated the mass of an ion pair to the mass of unit cell, but the unit cell only contains half an ion pair. The last line should read:
    $$\dfrac{58\cdot5}{L} = 2 \times 2\cdot16 \times (2\cdot8)^3 \times 10^{-24}$$

12. a, c, e
    H_2O, NH_3, and CH_4 each have four electron pairs around the central atom. Although not all these electrons may be taking part in covalent bonds the molecules will be tetrahedral.

13. a, c, e
    If some of the iron exists as Fe^{3+} there must be a number of cation holes. Thus there will be more oxygen atoms than iron atoms and the formula is more likely to be between $Fe_{0\cdot86}O$ and $Fe_{0\cdot92}O$.

14. d
    This is a very striking experiment. The most reasonable hypothesis is that the dipoles have all been orientated in the same direction and the negative ends have been attracted to the positively charged rod.

ANSWERS AND COMMENTS

15. a, b

The differences in the properties of ethanol and phenol appear to be due to the differences between the C_2H_5 and the C_6H_5 groups.

The boiling point of ethanol is much higher than that of ether in spite of the higher molecular weight of ether. The differences in surface tension and viscosity are also unlikely to be due only to the differences in dipole moments. Nitrobenzene has a very high dipole moment and when this is taken into consideration the values of the other physical properties do not appear to be unduly high.

16. b, d, e

Care must be taken in estimating dipole moments, because the value of the dipole gives no information as to which is the positive or negative end. The dipoles of phenol and nitrobenzene are in opposite directions, i.e. electrons are drawn from the benzene ring in the case of nitrobenzene and into the ring in the case of phenol. We would expect the boiling point of $C_6H_5C_2H_5$ to be higher than that of $C_6H_5CH_3$.

17. c, d, e

18. b, c, d

Lyophilic colloids are water-attracting and the process of precipitation is reversible. Lyophobic colloids are water-repellant and the process of precipitation is irreversible. Most metal colloids are negatively charged.

19. a, c, d

20. a, c, d, e

Hardy-Schulze rule: the precipitating effect of an ion on a disperse phase of opposite charge increases with the valency of the ion. We must conclude that the colloid X is negatively charged and that Y is positively charged. Hence is will take far less Al^{3+} to precipitate X. Similarly we can forecast that SO_4^{2-} ions will be better in precipitating Y and PO_4^{3-} ions will be better still.

REACTION KINETICS

Answers and comments

1. a, d, e

2. a, c
The volume of CO_2 is dependent upon the quantity of acid used according to the stoichiometry of:
$$2HCl + CaCO_3 \rightarrow CO_2 + H_2O + CaCl_2$$

3. a, c, d
4. a, c, d, e
5. a, c, d

6. b, d, e
The criteria for a first order reaction of the type Reactant (R) → Products being:
(i) $\dfrac{d[R]}{dt} = -k\,[R]$
(ii) log [R] plotted against t gives a straight line.
(iii) The half life of the reactant is independent of the concentration of the reactant.

7. a, c, d, e
8. c
9. a, c, d
10. a, b, c, d
11. a, b, c, d, e
12. a, b, d, e
13. d

14. b, e
The criteria for a second order reaction of the type Reactant (R) → Products (or of the type A + B → products, in which the reactants A and B always have the same concentration) being:

(i) $\dfrac{d\,[R]}{dt} = -k\,[R]^2$

(ii) $\dfrac{1}{[R]}$ plotted against t gives a straight line.

(iii) the half life depends on the initial concentration, $t_{\frac{1}{2}} = \dfrac{1}{k\,[R]}$

ANSWERS AND COMMENTS

15. a, b, c, d, e
16. a, b, c
17. c, d, e
18. b, c
19. b, c, d, e
20. a, d, e

When the count is $\frac{1}{2}$ the original count, $\log_e 2 = k\, t_{\frac{1}{2}}$
when the count is $\frac{2}{3}$ the original count, $\log_e \frac{3}{2} = k\, t_{\frac{2}{3}}$

On dividing these, we get $\dfrac{\log_e 2 \cdot 0}{\log_e 1 \cdot 5} = \dfrac{\log_{10} 2 \cdot 0}{\log_{10} 1 \cdot 5} = \dfrac{t_{\frac{1}{2}}}{t_{\frac{2}{3}}}$

$$t_{\frac{2}{3}} = \frac{5600 \times 0 \cdot 1761}{0 \cdot 3010} \text{ years}$$

$$t_{\frac{2}{3}} \simeq 4100 \text{ years}$$

PERIODIC TABLE

Answers and comments

1. c, d; b is meaningless

2. c, e
In view of the considerable number of natural isotopes and their wide range of abundances, it is surprising that the order of elements according to atomic numbers is so well matched by their atomic weights.

3. a, b, c, e
Care must be taken when dealing with radii, as non-metals are treated differently from metals (see Introduction section 15).

4. c, d, e

5. d, e
These are typical properties of the oxides and chlorides of non-metals. Carbon tetrachloride is unusual as it is covalent but is not hydrolysed by water.

6. a
As a period is crossed there is a change in the bonding of the hydrogen atom (i.e. from H^- through H covalent to H^+ when in aqueous solution) depending upon the electronegativity of the element with which it is combined.

7. a, b, c, d, e
These are the typical variations found in the groups.

8. a, b, d
The changes occurring down a group are caused essentially by the increase in atomic radius and consequent greater ease of ionization. The atomic volume (i.e. the volume occupied by 1 mole of the element calculated from the atomic weight and density) is a less accurate measure of atomic size, but is still useful.

9. a, b, c, d, e
Group numbers are less useful for transition metals, but give some indication of the valencies (or better, oxidation states) to be expected in their compounds.

10. c, e
The rare (noble?, inert?) gases are the key to an elementary explanation of bond formation. This is still useful even though some of these elements do form compounds such as XeF_4 and Na_4XeO_6.

11. c, d, e
The main energy levels (K, L, M, etc.) are subdivided, and the s and p levels together give the familiar 'octet'. The sub-levels are particularly useful in dealing with transition metals (d level filling) and rare earth metals (f level filling). Although a rare gas electron structure is

ANSWERS AND COMMENTS

produced in **b.**, far too much energy is required to give X^{4+} and four covalent bonds are formed instead.

12. a, b, d, e
These species are isoelectronic with Ar.

13. b
Energy is needed to form ions. M^+ would not be produced in chemical reactions as this ion has no electron stability.
There are in fact five stable isotopes of Ca.

14. b, c, d, e
The measured bond angles are:

BF_3	120°	Planar molecule
BF_4^-	109°	Tetrahedral ion
NH_3	107°	Slightly distorted tetrahedral molecules
H_2O	104°	

(See Introduction section 10.)

15. d, e

16. b

17. a, b, c, e
These are but a few of many such nuclear reactions. In each transformation the sum of numbers of protons and neutrons must remain constant. **c** is the reaction which gave the first artificially produced radioactive isotope.

18. a
Emission of an alpha particle (^4_2He) means that the element produced is displaced two groups to the left in the periodic table because of the loss of two protons.

19. a, b, c, d
It is impossible to predict whether a particular isotope will be radioactive, but the main factor seems to be the ratio of protons to neutrons in the nucleus.

20. e
The only hope of making such unstable nucleii is to 'shoot' into a more stable nucleus the number of protons needed, contained in one particle. **e** is the method which Russian scientists are said to have used in 1964.

GROUP 1 *M* THE ALKALI METALS

Answers and comments

1. b
The metallic radius is considered to be half the distance between the centres of atoms in the metal. Rubidium has the second largest metallic radius in the group Li to Cs.

2. d
Energy is always needed to ionize a metal atom. As the potassium atom has a larger radius than the sodium atom, less energy is needed.

3. a, b, c

4. c, e
The larger the radius of an atom the easier ionization, and hence reaction, occurs. The calcium atom is smaller than the caesium atom, but loses two electrons which requires more energy than the loss of one.

5. a
This ratio (the atomic volume) gives a readily calculated and useful idea of the relative sizes of atoms.

6. a, b, c, e
A sodium-lead alloy is essential in the manufacture of lead tetraethyl. Titanium is extracted by the reduction of titanium tetrachloride using sodium or (more generally) magnesium, as both are reducing agents under these conditions.

7. c
Sodium peroxide is made by the careful oxidation of sodium in air. The equations for this sequence of reactions show that 1 mole of Na_2O_2 liberates enough iodine to react with 2 moles of $Na_2S_2O_3$, i.e. 2000 ml of M $Na_2S_2O_3$.

8. e
Anhydrous sodium carbonate can only be made in this country by heating a hydrate above 100°C. In Lagos, washing soda deliquesces!

9. c

10. b, e
H^- only appears in hydrides of very readily ionized metals. Combination with hydrogen is usually classed as reduction, but here the change is $Na \rightarrow Na^+$ which is oxidation.

11. a, d

12. b
Hydrogen carbonates in 0·1 M solution are slightly alkaline because of hydrolysis.

ANSWERS AND COMMENTS

13. a, e

Sulphites and hydrogen sulphites should not give precipitates with acidified barium chloride solution. They sometimes give faint ones because of oxidation to sulphates. 'Harpic' is largely sodium hydrogen sulphate.

14. a, e

The brown fumes are nitrogen dioxide, not bromine.

15. a, b, c

16. a, b, c

17. a

The first element in a group often is unusual. The cause is the very small radius of both Li and Li^+.

18. a, b

Catalysts are rarely used in ionic reactions.
Sodium hydrogen carbonate decomposes quite rapidly at 80°C.

19. d, e

20. c

    a. The first salt to crystallize is sodium chloride (being the least soluble). This is the industrial method of preparing potassium nitrate.

    c. Common salt is commercially prepared by vacuum evaporation.

    d. The first part of the curve is that of $Na_2SO_4, 10H_2O$, which dehydrates at higher temperatures.

    e. The transition temperature is 33°C (from these figures).

GROUP 2 M

Answers and comments

1. b
2. d
3. b
4. b, c

As the group is descended, the atomic radii increase. This is mirrored in the changes in ionization energies and standard electrode potentials. These are not always comparable though, as electrode potentials refer to hydrated ions, whereas ionization energies do not.

As a period is crossed the atomic radii decrease slightly because of the higher nuclear charge attracting the electrons more strongly, so calcium atoms are smaller than potassium atoms.

5. a, c, e

Calcium hydroxide is one of the relatively few compounds whose solubility decreases with temperature rise.

c can be predicted using Le Chatelier's Principle.

Phenolphthalein and methyl orange both give the same end point, as the alkali is a strong one (even though the solution is rather dilute).

6. e

This is approximately the concentration of magnesium salts in sea water, from which the metal is extracted commercially.

7. c

Magnesium burned in air gives both the oxide and the nitride. Can you think of a way of showing this in the laboratory?

8. b, c, d, e

The production of anhydrous magnesium chloride for electrolysis is an essential step in the extraction of the metal from sea water.

a gives magnesium oxychloride because of hydrolysis.

9. c
10. a, b, d, e
11. a, b, d, e

The terms *carbonate* and *non-carbonate* for water hardness are better than *temporary* and *permanent*.

'Calgon' (sodium hexametaphosphate) is a very effective softening agent.

12. a, b

The 'certain oxides' include alumina and silica.

This important process gives cement and sulphuric acid from calcium sulphate, found mainly near Billingham.

ANSWERS AND COMMENTS

13. c, d, e
14. c, d, e
15. a
16. e

These questions illustrate the regular changes in solubilities of compounds of the metals in this group.

17. a, c

Because of hydrolysis, sodium carbonate solution contains an appreciable concentration of $OH^-_{(aq)}$, thus giving a basic salt. The concentration of $OH^-_{(aq)}$ is much less in sodium hydrogen carbonate solution.

18. a, e

The chemical properties of radium follow those of the rest of the group. In addition, of course, the metal and its compounds are radioactive.

19. a

Unexpected properties are sometimes shown by the element at the head of a group. The reason here is the very small ionic radius of Be^{2+} together with its large ionic charge.

20. a

In certain reactions lithium shows a marked resemblance to magnesium (the so-called diagonal relationship). The probable reason is that the radii of Li^+ and Mg^{2+} are almost identical.

GROUP 3 M

Answers and comments

1. e
Much energy is needed to produce Al^{3+}. Even more is needed to ionize the smaller boron atom and B^{3+} does not exist at all in compounds.

2. c
Three electrons must be removed from each Al atom, so the standard electrode potential is less negative.

3. a
Because of the impossibly high ionization energy, boron is a typical non-metal forming covalent compounds.

4. c, d

5. a, b, e
The covalent compounds BCl_3 and $AlCl_3$ are electron deficient (six electrons with each B and Al atom). The simple compounds therefore tend to polymerize (as in Al_2Cl_6) or form addition compounds with molecules containing lone pairs of electrons

as in

$$\begin{array}{c} H \\ \diagdown \\ H-N: \rightarrow Al-Cl \\ \diagup \quad \diagdown \\ H \qquad Cl \end{array} \begin{array}{c} Cl \\ \diagup \\ \\ \\ \end{array}$$

6. b, c, e
It would be expected that as the group is descended, the elements will become more metallic, the oxides more basic, and so on. This is generally the case but there are exceptions, e.g. **a** and the fact that thallium nitrate is $TlNO_3$.

7. a, c, d
The high affinity of aluminium for oxygen is the basis of the 'Thermit' and other similar reactions.
On energetic grounds it is likely that reaction **d** will take place as 3260 kJ are evolved when aluminium combines with 3 moles of oxygen and only 2230 absorbed when 2 moles of chromium trioxide are thermally decomposed. Thus although a reaction between aluminium and chromium trioxide might be expected, nothing can be said from this about the rate at which it may happen. (In this case it is fast enough to be commercially profitable.)

8. c, d, e

9. c, d, e
Although aluminium is high in the electrochemical series it reacts less vigorously than expected

with air and acids because of a very thin film of surface oxide. This can be dissolved by alkalis and hydrochloric acid, but not by oxidizing agents such as concentrated nitric acid.

10. a, c, d

11. a, c, e

Alums are all isomorphous, but not all monovalent or trivalent metals form them as ionic size is the determining factor. It is debatable whether a mole of aluminium potassium sulphate is K_2SO_4, $Al_2(SO_4)_3$, $24H_2O$ or $K Al(SO_4)_2$, $12H_2O$.

12. b, e

A scheme similar to this is used industrially to obtain pure alumina from bauxite which often contains ferric oxide and silica impurities.

13. b, d

Great care is taken to purify the ore before the aluminium is extracted, as impurities (especially iron) cause rapid corrosion.

14. b, d

Copper, being a transition element, forms complexes with ammonia, whereas aluminium does not.

15. c, d, e

Al^{3+} in solution is hydrated with probably (on average) 6 H_2O per ion. Because of the very small radius of the ion and its large charge, the electron arrangement in the water molecules is altered and hydrogen ions are produced making the solution acidic.

i.e. $[Al\ 6H_2O]^{3+} \rightleftharpoons [Al\ 5H_2O.OH]^{2+} + H^+(aq)$

a and **b** are at first sight correct, but it is highly unlikely that Al_2S_3 and $Al_2(CO_3)_3$ are produced even momentarily. The concentration of $H^+(aq)$ thus produced by hydrolysis in aluminium sulphate solution is greater than that given by H_2S or CO_2 so these gases are liberated and hydrated $Al(OH)_3$ precipitates.

16. b, c, d

Cations having a large charge and small radius will give acidic aqueous solutions (see comment on previous question).

17. a

The remainder give the metal hydroxide or a basic salt by hydrolysis.

18. a, e

Aluminium bromide may be formulated as $Al^{3+}\ 3Br^-$, but the tendency of Al^{3+} to distort the electron clouds of the larger Br^- is considerable and for practical purposes the compound behaves as a covalent one. There are relatively few metal compounds which can be considered to consist almost entirely of ions (e.g. Cs^+ F^-). Most exhibit partial covalent properties to a greater or lesser extent.

19. b, c, d
See comments on Questions 5 and 18.

20. a, b, c, d, e
These were exactly the predictions made by Mendeleeff, and were found to be accurate when gallium was isolated in 1875.

GROUP 4 M

Answers and comments

As the group is descended metallic character appears, as is to be expected from the increasing atomic radii. Carbon and silicon are non-metals. Germanium is more metallic than non-metallic (see the 'steps' in the periodic table) but shows both characteristics: e.g. GeO is amphoteric and GeH_2 is thermally stable up to 285°C. Tin and lead are progressively more metallic.

1. a, b, c
These values indicate that although C and Si do not form ionic compounds, Ge might be expected to. Ge^{2+} is formed in solution as are Sn^{2+} and Pb^{2+}.

2. b
The values are generally in the expected sequence, but not always (e.g. the rather low melting point and ionization potential of Sn).

3. c, d, e
Ge, Sn and Pb have oxidation states of $+2$ and $+4$, but it is unlikely that M^{4+} is formed in compounds as too much energy is required.

4. a, c, d
$PbCl_2$ is insoluble in cold water.

5. b, c, d, e
Covalent chlorides are almost invariably hydrolysed by water. This reaction substitutes —OH for the more electronegative halogen atom. CCl_4 is unusual in this respect.

6. b, e
Bond energies are a measure of the energy needed to split a bond. These values agree with experimental fact: e.g. the reluctance of Si to form chains and the great thermal stability of silicates.

7. c
As the periodic table is crossed the hydrides change from salt-like compounds (Na^+H^-) through covalent ones (CH_4) to hydrides which are acidic in solution ($H^+(aq)Cl^-(aq)$). The electron deficient molecule BH_3 is unknown, the simplest hydride of boron being B_2H_6.

8. a, c, d, e
Far too much energy is needed to ionize these, even if the ion so formed has a rare gas electron structure.

9. d
Magnesium silicide with acid gives a mixture of spontaneously flammable silanes (mainly SiH_4 and Si_2H_6).

ANSWERS AND COMMENTS

10. b, d, e
These have the covalent structure of diamond and are all extremely hard.

11. b, c, d, e
The electron structure of CO is such that the molecule has lone pairs of electrons which form strong bonds with transition metals (e.g. iron in haemoglobin).

12. d, e
Siloxanes when polymerized give the widely used silicones. Germanium is a constituent of transistors. Lead is a cumulative poison and lead chloride and carbonate are not completely insoluble in water.

13. a, c, d, e
The present-day high cost of tin has led to efforts to reclaim it from cans via the tetrachloride.

14. b
Tin(II) in solution is readily oxidized to tin(IV). This does not apply to lead(II).

15. a, b, d
Certain silicates with their three-dimensional arrangement of Si and O atoms contain holes of molecular size which will trap (say) butane but allow propane, a smaller molecule, to pass. Such molecular sieves are proving most useful in many applications where separation of chemically similar molecules is required.

16. c, d, e
Plumbous plumbate ($Pb_2^{II}[Pb^{IV}O_4]$) is the best designation of red lead.

17. a, b, c
Calcium plumbate is one of several pigments with good anti-corrosion properties.

18. c, e
Aluminium and tin(IV) chlorides are much more covalent than ionic in character and are therefore readily hydrolysed by water.

19. b, c
Mercury(I) is rather unusual and appears not to exist as the simple Hg^+ ion.

20. a, c, d
Mendeleeff accurately predicted the properties of '*eka*-silicon' several years before it was discovered.

GROUP 5 M

Answers and comments

As this group is descended the elements become progressively more metallic, although Bi still has some non-metallic characteristics (see the 'steps' on the periodic table).
Decreases in ionization energies and electronegativity values will be seen with increasing atomic radius.

1. a
No ionization is expected for N and P, but Bi^{3+} does exist in solution. There are no important **salts** of As or Sb.

2. c
This is easily seen if the values are put in the periodic table.

3. e
The oxidation state of X is $+3$ in each case.

4. a, b, c, d, e
All the values given are in fact correct, although **b** and **e** might be expected to be different. In general there are regular changes down a group, but values are sometimes anomalous.

5. b, d, e
Most M elements forming amphoteric oxides are situated near the 'steps'.

6. b, e
Elements low down in groups 3 or 5 form salts of this type.
This is expected more in group 3 than in group 5.

7. d
The halides of As, Sb and Bi clearly show the change from non-metallic to metallic characteristics. The conductivities for the molten compounds are:
$AsCl_3$ 1.2×10^{-6} $ohm^{-1} cm^{-1}$
$BiCl_3$ 4.1×10^{-1} $ohm^{-1} cm^{-1}$

8. c, d, e
This is the basis of the famous Marsh Test for arsenic. Covalent hydrides of elements low down in groups 4, 5 and 6 are thermally unstable.

9. c, d, e
Hydrolysis to the oxychloride (XOCl) or hydrated oxide occurs readily. NCl_3 is violently explosive.

10. a, b, c
The reaction between hydrogen and nitrogen is an equilibrium one, but it is doubtful whether equilibrium is reached in industry. The conditions are usually 200 to 500 atmospheres and

350 to 400°C (with a catayst) lwhen an overall conversion of 10 to 20 per cent occurs. When a mixture of 1 mole of nitrogen and 3 moles of hydrogen reaches equilibrium at 1000 atmospheres and 200°C about 97 per cent conversion occurs, but this is the **maximum** rather than the **optimum** conversion needed industrially.

11. a, b, c, d, e
The major uses of ammonia are as shown (U.K., 1959).

Use	percentage of total ammonia produced
Fertilizers	75
Chemicals	7
Explosives	5
Fibres, plastics and resins	6
Others	7

12. c, d, e
Acids are conveniently defined as proton donors, and bases as proton acceptors (Brønsted). In non-aqueous solvents this is still applicable. NH_4^+ can lose a proton to become NH_3 (the solvent) and H^+, and thus act as an acid.
c is therefore neutralisation of an acid by a base.
d is analogous to this reaction in aqueous solution:
$$Zn(OH)_2(s) + 2KOH(aq) \rightarrow K_2[Zn(OH)_4](aq)$$
e is analogous to this reaction:
$$Ca(s) + 2H^+(aq) \rightarrow Ca^{2+}(aq) + H_2(g)$$
There are many such reactions in a variety of other non-aqueous solvents.

13. a, c, d
The electron structure of this molecule enables it to form many complexes of the type $[FeNO]^{2+}$, which is the cause of the brown colour in the brown ring test.

14. a, b, e
NO_2^- in acidic solution can act as an oxidizing or reducing agent depending on the other reactant.
The order of oxidizing power here is $Br_2 > NO_2^- > I_2$.
The formal equation can be written:
$$Zn(s) + 4HNO_3(aq) \rightarrow Zn(NO_3)_2(aq) + 2NO_2(g) + 2H_2O(l)$$
but this is far from the truth, as many other reduction products of nitric acid are formed.

15. d
The reaction between non-metals and alkalis often involves disproportionation.

ANSWERS AND COMMENTS

	Reaction	Oxidation states of non-metal
e.g. P + OH⁻	P → PH₃	$3e^- + P^° \rightarrow P^{-III}$
	P → H₂PO₂⁻	$P^° \rightarrow P^{+I} + e^-$
Cl₂ + OH⁻	Cl → Cl⁻	$e^- + Cl^° \rightarrow Cl^{-I}$
	Cl → ClO⁻	$Cl^° \rightarrow Cl^{+I} + e^-$

16. b, d, e

Phosphine prepared as above may easily be collected and handled if first passed through kerosene to remove spontaneously flammable P_2H_4.

17. a, b, c, d

In each of these the oxidation state of nitrogen increases from N^{-III} to $N^°$ (i.e. $N_2(g)$) and therefore the other reactant is reduced.

18. a, b

a. N^{+I} would be expected to be oxidized to N^{+V} (NO_3^-).
b. P^{+III} would be expected to be oxidized to P^{+V}(PO_4^{3-}).
c and d. As^{+V} and Sb^{+V} are not likely to be oxidized as +5 is the highest oxidation state in this group.
e. Bi^{+III} is unlikely to be oxidized to Bi^{+V}.

19. b, c

This is most easily seen using oxidation states as in Question 18.

20. a, d

These may be the predicted properties of an element occurring below Bi, but Pa is never placed in this group in modern tables as it is an actinide giving quite different reactions.

GROUP 6 M

Answers and comments

As this group is descended, atomic radii increase (as do atomic volumes which give some idea of relative sizes of atoms) and the tendency to form positive ions also increases. These fairly regular changes are to be expected and are shown in most of the data given, although strict regularity is not always observed: e.g. the electron affinity for Se is higher than expected. With Te and Po some metallic properties are expected and observed, as in the formation of basic sulphates and nitrates.

1. b
2. e (See Introduction section 7)
3. b
4. d, e
5. b, e

The allotropy (polymorphism) of sulphur and the variety of forms in which it exists all arise from the breakdown of S_8 rings as the temperature rises, and the consequent formation of chains of varying length.

6. c
7. e

All these are possible ways of preparing oxygen in the laboratory, but e is the most convenient.

8. b, c

Hydrogen bonds (which are believed to be mainly electrostatic in nature) account amongst other things for many of the properties of that unusual compound, water.

9. d, e

As a period is crossed the hydrides of elements alter from ionic salt-like compounds (Na^+H^-) through covalent ones (CH_4) to compounds where electron sharing is uneven (H_2S, HCl). The latter ionize to $H^+(aq)$ in water. The difference in electronegativity between the element and hydrogen is a key factor.

10. b

Oxidation states		
+2	Cu^{2+}	A change in oxidation state of an element to a lower
	↑	and to a higher value at the same time is conveniently
+1	Cu^+	termed disproportionation.
	↓	
0	—Cu—O_2—	
	↑	
−1	H_2O_2	
	↓	
−2	H_2O	

ANSWERS AND COMMENTS

11. a, d, e   See comment on Question 12.

12. a, b, d, e

Oxidation and reduction are relative terms. SO_2 is usually classed as a reducing agent but in these circumstances H_2S is a better reducing agent.

13. a, b, c, e

It is difficult to apply the principles relating to equilibria in closed vessels (i.e. Le Chatelier's Principle and the Law of Equilibrium) to the industrial flow processes where equilibria are rarely reached. In this case most of the conditions which increase the yield of SO_3 are applied industrially, where a 95 per cent overall conversion of SO_2 to SO_3 is quoted. These conditions, however, are those for **optimum** rather than **maximum** conversion. The same applies in the Haber process.

14. b, c

The chamber process is not yet dead in the United Kingdom, although more recent versions use packed towers rather than chambers. It has the advantage of working on impure sulphur dioxide and is in wide use in the U.S.A.

15. a, b, c

i.e. those compounds in which the oxidation state of sulphur is less than $+6$.

16. a, c, d, e

The 'resting state' of sulphur in its oxy-acids is $+6$ (as in SO_4^{2-}). Oxy-acids in which the oxidation state is less than this might be expected to reduce iodine or potassium permanganate. $Na_2S_2O_6$ (sodium dithionite) apparently does not!

17. a, b, c, d, e

Selenium is obtainable from chemical suppliers. Enough to make iron(II) selenide (by heating with iron filings) can be purchased for a few shillings and H_2Se readily prepared. It is more poisonous than H_2S.

18. a, b, c, e

Although the element has a metallic sheen, selenium is essentially non-metallic in its chemical properties.

19. b, c, d, e

All these properties show Te to be non-metallic, but its position in the periodic table would indicate some metallic character. (It does form a basic salt $Te_2O_3(OH)_3NO_3$.) Unlike SO_2, TeO_2 is insoluble in water although it is soluble in sodium hydroxide solution to give Na_2TeO_3.

20. a, b, c, d, e

Polonium is in many ways metallic, as expected. The element and its compounds are, of course, radioactive.

GROUP 7 M THE HALOGENS

Answers and comments

1. e
This can easily be seen if the values are inserted in the periodic table. F is the least likely of the halogens to form X^+ in compounds and the most likely to form X^-. The properties of the group are centred around this.
(It should be noted that xenon needs less energy to ionize than oxygen. This was one of the facts which encouraged the successful attempts to prepare compounds of the 'inert' gases.)

2. a
Fluorine is the best oxidizing agent known and it can only be made by the electrolytic oxidation of F^- in the absence of water (which it oxidizes).

3. a might be expected to be correct, but in fact the values are CH_3F 1·81: CH_3Cl 1·83: CH_3Br 1·79: CH_3I 1·64 (Debye units). **e** is symmetrical with no dipole moment.

4. c, d, e
This again emphasizes that fluorine is a stronger electron acceptor than the other halogens.

5. b, c
All are possible methods of chlorine production (i.e. $Cl^- \rightarrow Cl$) but **d** and **e** are the industrial processes.

6. b
This is the same for cold dilute and hot concentrated sodium hydroxide solutions, even though the products differ.

7. c, d
The first product is the hydrogen halide, but HBr and HI are readily oxidized by conc. sulphuric acid, whereas HF and HCl are not. HBr and HI must be prepared in the laboratory by another method.

8. d
Chlorine is a better electron acceptor than bromine (i.e. is more electronegative) and the uneven electron sharing is shown by the dipole moments of these covalent compounds.

9. a, b, c
HF is unusual in this respect, being not fully ionized in solution.

10. c

11. c, d

12. c, d

ANSWERS AND COMMENTS

13. a, b

It is reasonable to suppose that **13 b** would happen, but the reaction is too slow. In the redox potential table the tendency for the reaction from left to right to occur becomes greater as the redox potential becomes more positive. This is a continuation of the familiar reactivity or electrochemical series for metals. It should be noted, however, that the numerical values are relevant only if the reactions take place under the conditions used to measure the values (i.e. molar solutions of all ions) and that a reaction which is theoretically possible may proceed too slowly to be of use (see Question 14).

14. b, d

Oxidation of Br^- can theoretically be done using oxygen, but this is too slow. A large plant is sited at Amlwch, Anglesey, for extracting bromine from acidified sea water using chlorine.

15. a, b, c, d

As iodine is at the bottom of the group, $I^- \rightarrow I$ occurs much more readily than $Cl^- \rightarrow Cl$.

16. c, d, e

Oxidation of these must result in reduction of nitric acid (N^V) to brown nitrogen dioxide (N^{IV}).

17. a, b

This can be used as a test for F^-. No other hydrogen halide attacks glass in this way.

18. a

$I^- \rightarrow I$ occurs readily and even Cu^{2+} is a good enough oxidizing agent to liberate iodine from potassium iodide.

19. d, e

Fluorine is not a typical halogen and has many unexpected properties.

20. a, b, d, e

It is reasonable to predict these properties, but the chemistry of astatine is as yet little known.

TRANSITION METALS

Answers and comments

In transition metals the d electron level is filling (there being a maximum of ten such elements in each period). This gives rise to the following properties which are shared by most transition metals to different extents.

   i. Variable valency (oxidation state).
   ii. Formation of complex ions.
   iii. Formation of coloured ions in solution.
   iv. Catalytic activity.

In rare earth elements, the f level is filling (there being therefore fourteen such elements in a period). They are sometimes termed *inner transition elements*.

1. b, e
The elements are (in order) Al, V, Na, Ne, Ti.
The lower values and smaller differences between values characterise transition elements. Although ionization energies are not measured in solution it is reasonable to expect element 3 to ionize in solution to M^+ only, whereas element 2 might be expected to form M^+ through to M^{5+}, or at least have oxidation states $+1$ to $+5$. (V has oxidation states V^{II} to V^V.)

2. a, b
Here the $3d$ level is filling.

3. c
Here the $4f$ level is filling.

4. a, d, e
In a transition element there is an unfilled d energy level. In Zn, Cd, and Hg this level is full, so are these transition metals? This can be debated as the evidence is not conclusive in either direction.

5. a, b, c, d
Other elements have some of the properties normally associated with transition metals.

6. a, b, c, d, e
In a complex ion (or molecule) the central atom (usually a transition metal) is surrounded by and attached to electron donating groups. The effect is to destroy the properties normally associated with the central element.
   e.g. $[Fe(CN)_6]^{3-}$ has none of the properties associated with Fe^{3+}.

7. a, c, d, e
Only groups with unused 'lone pairs' of electrons can act as ligands. Water is weakly bound in most hydrated ions and is easily replaced by other groups.
   e.g. $[Cu4H_2O]^{2+} \rightarrow [Cu4NH_3]^{2+}$
This is done by adding $NH_3(aq)$ to copper(II) sulphate solution. (see note, page 19)

ANSWERS AND COMMENTS

8. e
All except **e** are double salts, not complexes.

9. a, b, c, d

10. b, e
The term valency loses its meaning with complexes. Oxidation states are much more informative.

11. a, b
An oxidizing agent is necessary to convert Cr^{III} to Cr^{VI}.

12. d
The oxidation state of Cr in CrO_4^{2-} and $Cr_2O_7^{2-}$ is the same (i.e. Cr^{VI}).

13. c
Any acid can be used here, but if relatively pure potassium permanganate crystals are needed, the other potassium salt produced needs to be very soluble in water so that it does not crystallise. Potassium carbonate is very soluble, but most of the other common potassium salts are less soluble.

14. a, e
A bond formed between Fe and CN^- means that the normal properties of both are altered.

15. c

16. b
b is a double salt giving $Cu^{2+}(aq)$: the rest are complexes.

17. a, b, c
As these form complexes with Ag^+, all should work. Potassium cyanide was used at one time until the much safer 'hypo' was introduced.

18. b, d, e
In many complexes, the central atom does not have a rare gas electron structure. Molecules such as ethylene diamine form chelates. (Greek: *chele* = claw)

19. c
1 mole of compound gives 4 moles of $H^+(aq)$ (i.e. 2 moles of H_2SO_4). The only salt to do this is **c**.
Such a method can be used to estimate copper and other metal salts in solution.

20. a, b, e
b is more instructive than **a**, as 1 mole of the compound gives 2 moles of chloride ions in solution.

ALKANES, ALKENES, AND ALKYNES

Answers and comments

1. b, e
Decarboxylation reactions of this type are of laboratory interest only and give impure products.

2. e
This is the Kolbé alkane synthesis: of limited application.

3. a, c, d
Industrially, the reverse process is more valuable (e.g. ethanol from ethylene).

4. b, c
Butylene is vague: but-1-ene is unambiguous.

5. b, d, e
Ethylene is prepared and used industrially in huge quantities, mainly for polythene.

6. a, d, e
Care is needed in handling this useful but thermally unstable compound. It is now increasingly synthesized from oil feedstock.

7. b
This type of reaction never gives pure products.

8. b, e
The main product is predictable from modern electronic theory or more empirically by Markownikoff's Rule.

9. a, c, d
Pure products are rarely obtained even in addition reactions.

10. a, d, e
d and e are important industrial applications of this reaction. 'Teepol' is the sodium alkyl sulphate derived from C_8—C_{17} alkenes.

11. a, b, c, d
This is the key reaction for the production of margarine.

12. a, b, d

13. a, b, d, e
The simple laboratory reactions of the carbon-carbon double bond are faster than those of the treble bond, which is rather unexpected.

14. a, b, c, d, e
All are polymerized ethylene or substituted ethylenes.

ANSWERS AND COMMENTS

15. b, d

d could be but-1,3-diene.

16. a, d, e

An H atom attached to the C atom of a treble bond can be replaced by silver (and other metals) to give thermally unstable acetylides.

C_4H_6 could be any of these structural isomers
  i. $CH_2=CH—CH=CH_2$
  ii. $CH_3—C\equiv C—CH_3$
  iii. $CH_3—CH_2—C\equiv CH$

17. a, b, e

Benzene (C_6H_6) does not decolourise these solutions even though there are apparently double bonds present.

18. c

This is the only saturated compound present.

19. b, c, d

These are the reactions expected of propylene, industrially a very important alkene.

20. a, b, c, d, e

The properties of double (and treble) carbon-carbon bonds arise because of the electron distribution between the atoms.

ALCOHOLS AND ALKYL HALIDES

Answers and comments

1. a, c
Hydrolysis is best done by refluxing with sodium hydroxide solution. Acids are very resistant to reduction, although lithium aluminium hydride (LiAlH$_4$) is quoted as giving good yields.

2. a, b, c, d
This is a very useful test for the —OH group in compounds, which must, of course, be free of water.

3. a, c, d, e
The answers show the sequence which is the most convenient and useful way of doing such calculations. All quantities used in preparations should be converted to moles.

4. a, b
These reagents are convenient to test for —OH groups in alcohols (provided the alcohol is dry), but the reactions are not always straightforward and other compounds in addition to the alkyl chloride are often formed.

5. a
This is one route to acetone, using either conc. sulphuric acid or by direct hydration with a catalyst, and then oxidizing the alcohol produced.

6. c
The first product is acetaldehyde which boils off before it is further oxidized. It is convenient to write the ionic half equation:
$C_2H_5OH(l) \rightarrow CH_3CHO(l) + 2H^+(aq) + 2e^-$
the electrons being taken up by $Cr_2O_7^{2-}$ in acid solution (see work on reduction-oxidation reactions).

7. d
Under reflux, acetaldehyde (b.p. 21°C) would be returned to the reaction mixture and be oxidized to acetic acid.
i.e. $H_2O(l) + C_2H_5OH(l) \rightarrow CH_3COOH(l) + 4H^+(aq) + 4e^-$

8. e

9. a

10. e
Tertiary alcohols resist oxidation, or are broken down completely to carbon dioxide, water and acids.

11. a, d, e
Glycerine is also widely made from propylene and used on a large scale for alkyd resins, drugs, cellophane, explosives.

ANSWERS AND COMMENTS

12. a, b, d, e
The iodoform reaction is NOT a specific test for ethanol. It works for any simple compound with the (CH₃CHOH—) or (CH₃CO—) group except acids and their derivatives.

13. c, d, e
Alkyl iodides are readily hydrolysed by aqueous alkalis.

14. b, c, d, e
Aqueous and alcoholic alkalis do give different products with alkyl halides, but the latter reaction is not of great importance. Ethanol is used to dissolve both the KCN and the alkyl halide, giving a homogeneous mixture. There are a number of ways in which organic reductions may be carried out. A widely used method involves the reaction of metals with an acid or alcohol (e.g. sodium + alcohols, zinc + acids, metal couples + alcohols) and these are most conveniently termed dissolving metal reductions.

15. a, c, d
Quantities should always be converted to moles and compared with the theoretical equation. This will show which reagent is in excess (in this case there is excess P and C_2H_5OH).

16. a, b, d, e
The percentage yield must be calculated from the number of moles of reactant which is NOT in excess (i.e. propanol).

17. a, b, c, d, e
Different products (ether or ethylene) can be obtained from ethanol by altering the conditions. This is true of many reactions, particularly those involving catalysts.

18. a, b, d, e
The two functional groups in bromoethanol behave as expected of alcohols and alkyl bromides. The compound is readily made by bubbling ethylene into bromine water. It boils at 146°C.

19. a, c, d, e
Grignard reagents such as this (CH_3MgI) are very useful in a wide variety of syntheses.

20. a, c, d
Industrial practice varies widely and alters rapidly. Trends today are towards a petroleum-based industry (e.g. in **e** the ratio is 60:20:1). Reactions are as direct as possible and often divorced from the simple stages usually discussed in schools.
An example of this is the production of acetaldehyde.

Old method: $C_2H_5OH(l) + \tfrac{1}{2}O_2(g) \xrightarrow{\text{catalyst}} CH_3CHO(l) + H_2O(l)$

New method: (dehydrogenation) $C_2H_5OH(l) \xrightarrow[\text{300°C}]{\text{catalyst}} CH_3CHO(l) + H_2(g)$

The hydrogen can be used in many ways; e.g. the hydrogenation of vegetable oils to give margarine.

ALDEHYDES, KETONES, ACIDS, AND ESTERS

Answers and comments

1. c
The ionic half equation can be written
$$CH_3CH_2CH(OH)CH_3(l) \rightarrow CH_3CH_2COCH_3(l) + 2H^+(aq) + 2e^-$$

2. b
Butanone (a useful paint solvent) is resistant to further oxidation, as are most ketones.

3. d
A water condenser will stop the loss of acetaldehyde.

4. b, e
To be sure a compound is an aldehyde and not a ketone, both Fehling's and Tollens' reagents should give positive results.

5. c, d, e
This is a very convenient laboratory method to distinguish between primary, secondary, and tertiary alcohols.

6. a

The initial product might be e but
$$\begin{array}{c} H \quad OH \\ \diagdown \diagup \\ C \\ \diagup \diagdown \\ R \quad OH \end{array}$$
cannot be isolated and dehydrates to
$$\begin{array}{c} H \\ \diagdown \\ C=O \\ \diagup \\ R \end{array}$$

7. a, b, e
Aldehydes and ketones are unsaturated and therefore addition reactions occur with HCN, NH_3, and $NaHSO_3$. HCN addition gives an extra C atom in the molecule, but reactions involving NH_3 and $NaHSO_3$ are obscure and of little use. e is more usually termed reduction.

8. a
The reduction of a carbonyl group can be done with sodium and alcohol, zinc and acid (i.e. dissolving metal reductions), or more often catalytically.

9. a, e
Although the carboxyl (—COOH) group on paper is made up of the $>C=O$ and —O—H groups, an electron rearrangement occurs. The typical condensation properties of $>C=O$ disappear, and —H ionizes somewhat in water. If however these groups occur separately in the same molecule [as in c] then their individual properties are preserved.

10. d
Salts are best made by adding a metal carbonate to the acid.

ANSWERS AND COMMENTS

11. b, c, e

Oleic acid ($C_{17}H_{33}COOH$) is an unsaturated liquid while stearic acid ($C_{17}H_{35}COOH$) is a saturated solid.

12. a

c has a molar mass of 104, but is not dibasic.

13. a, b, c, d, e

In laboratory practice only $SOCl_2$, PCl_3, or PCl_5 are normally used.
$POCl_3$ and SO_2Cl_2 react with water and would be expected to convert an acid to its acid chloride. They are used in this way industrially, but with the sodium salt of the acid.

14. a, b, c, d

e (stearic acid) is a solid and would need to be refluxed with the alcohol and a catalyst before an ester would form.

15. b, c, e

Quantities in moles are useful: quantities by weight are much less so.

16. e

Distillation of a mixture of organic compounds before chemical separation and drying often results in the formation of such azeotropes (constant boiling mixtures).

17. a, b, d

The hydrolysis of **a** is a major industrial activity in the production of soap. **b** is used in cosmetics, **c** is a salt in toothpaste (Gibbs S.R.), **d** is a polyester, but **e** is a polyamide.

18. a, b, d

This is a very important application of Le Chatelier's Principle. A catalyst will reduce the time taken to reach equilibrium but has no effect on the position of the equilibrium. A combination of **a** or **b** with **d** and **e** would give the largest percentage yield in the shortest time.

19. d

$C_4H_6O_2$ is methacrylic acid (important in the manufacture of perspex) and this is one route used to make it industrially.

20. b, e

30 per cent of the rubber part of a tyre is carbon black. Formaldehyde is very widely used for plastics (e.g. urea-formaldehyde). Tri-acetylated cellulose is a useful material for fabrics (e.g. tricel).

AMINES, AMIDES, AND CYANIDES

Answers and comments

1. c
Hofmann's bromination (degradation) is sometimes useful in removing a carbon atom from a compound.

2. d
A useful method of adding a carbon atom to a compound is by way of the cyanide (made usually from KCN and the alkyl iodide).

3. b
Hydrogen and a catalyst are used.

4. a, b, c
Amines, being derived from ammonia, are basic.
Their salts are best designated as in **b**.

5. b
This is a useful way to distinguish between primary amines which give nitrogen, secondary amines which give the yellow nitrosamine oil, and tertiary amines (no reaction).

6. a, b, c, d
Acetylation of amines is a useful way of protecting the rather easily oxidized —NH_2 group. The original amine is readily obtained by hydrolysis.

7. e
Reactions such as this always give a mixture, but **e** predominates.

8. a, c, e

9. b
This preparation is a good example of the application of Le Chatelier's Principle. In **e** the total weight of acetamide produced might increase, but the percentage yield is unaltered.

10. a, c, e
Ammonium salts give ammonia readily on warming with sodium hydroxide solution, whereas amides and cyanides require boiling.

11. b, c, e

12. b, c
Alkyl halides need to be heated with ammonia under pressure before reaction occurs. The products are amines, which are not hydrolysed with alkalis.

13. b

14. b, e
KCN is usually dissolved in alcohol rather than water so that the mixture is homogeneous.

ANSWERS AND COMMENTS

15. c

16. b, e

17. b

The reaction between nitrous acid and primary amines always liberates nitrogen, but the yield of the alcohol varies widely. With methylamine no methanol is obtained, and ethanol from ethylamine is produced in low yield.

18. a, b, c, d, e

The two functional groups are well separated and behave independently. As one group is basic and the other acidic, the compound can form an internal salt as in **c** [a *zwitterion*]. Proteins are polymers of amino acids similar to this.

19. a, b, c, d

Although **a** and **e** are isomers, the position of the —NH_2 and —Cl groups in the molecule makes a great deal of difference to their chemical properties.
Mono, di, and trichloroacetic acids are progressively more fully ionized in water than the parent acid because of the electron attracting power of —Cl.

20. d

The only nitrogen-containing materials are nylon and Courtelle. The latter with free —CN groups gives ammonia, but nylon which is a polyamide does not.

AROMATIC COMPOUNDS

Answers and comments

1. a, c

2. a
Benzene reluctantly takes part in addition reactions. When these occur (the only important ones are with chlorine, bromine and hydrogen) the compound formed is a cyclic aliphatic one, and is not aromatic.

3. a, b, e
Benzene is generally unreactive and few groups can usually be directly substituted into the ring, namely —Cl or —Br, —CH$_3$ or —COCH$_3$, —NO$_2$ and —SO$_2$OH.

4. a
This is one of the few addition reactions of benzene. Benzene hexachloride has nine possible spatial isomers, but only the *gamma* form is an efficient insecticide (Gammexane). It behaves as a cyclic alkyl halide.

5. a, c, e
Substitution occurs with cold benzene and a catalyst (e.g. iron, iodine or other halogen carriers). The reaction can be stopped after the addition of each halogen atom by noting the weight increase. Pure products are rarely obtained and in this case *ortho* or *para* dichlorobenzene would be expected as impurities because a halogen atom in the ring is *o* and *p* directing (see later).

6. b
Mononitrobenzene is the main product. The —NO$_2$ group is *meta* directing and slows down further substitution. Molecules which can be formally written with several alternating double and single bonds are often coloured (in this case pale yellow).

7. a, e
Sulphonation generally occurs slowly and it is difficult to substitute a second group. The compound so produced is a monobasic acid. Sulphonation of an alkyl benzene with long side chains gives (after neutralization) a very effective basis for a number of synthetic detergents (Tide, Daz, etc.).

8. b, c
The Friedel-Crafts reaction (using AlCl$_3$, BCl$_3$ or other similar catalysts) is a useful way of adding an alkyl or acyl side chain. When such compounds are oxidized the side chain goes leaving benzoic acid.

9. b
Under these conditions (no catalyst and ultra-violet light) the side chain is chlorinated.

ANSWERS AND COMMENTS

10. c
A halogen atom attached to the ring is replaced only with considerable difficulty, whereas side chain halogens behave as alkyl halides.

11. a, b, d, e
In **b** the compound is hydrolysed to benzoic acid.

12. a, b, d

13. a, b, c, d
e gives phenol.
Only —NH_2 groups attached to the ring are diazotized. The diazo reaction is useful in various syntheses: e.g. iodobenzene and phenol are difficult to make directly from benzene.

14. a
The —OH group is modified by the ring, becoming measurably acidic. Consequently the ring is altered by the —OH group and substitution occurs far more readily than in benzene.

15. a, b, d
Electron attracting groups encourage the tendency for —OH in phenol to ionize. **d** is quite a strong acid (picric acid).

16. d
As in the previous question, electron-attracting groups reduce the availability of the lone pair of electrons on —NH_2 and this is seen in the lower dissociation constant.

17. b, d
Electron-donating groups activate the ring.
Styrene will decolourize bromine water, but this would be an addition reaction of the double bond in the side chain.

18. b, c
All these compounds contain the $>C=O$ group but because of electron rearrangement, this no longer keeps its identity in amides and acids (or their derivatives).

19. c, e
Benzene is not the only compound to show aromatic properties. Apart from heterocyclics like pyridine, there are many five- and seven-membered rings known which behave like benzene. All however are planar and have at least six 'spare' electrons spread around the ring.

20. a, b, c, d, e

AROMATIC AND ALIPHATIC COMPOUNDS

Answers and comments

1. a, c, d, e
Halogen atoms in a molecule are usually readily hydrolysable, except when attached directly to the benzene ring.

2. a, b, c, d, e
Hydrogen in —OH is not ionized in alcohols, but where the —OH group is attached to electron attracting groups such as $>C=O$, [benzene ring with $-NO_2$], $>SO_2$, then ionization occurs in water. The extent of ionization depends on the group: e.g. C_6H_5OH is a weak acid, whereas $C_6H_5SO_2OH$ is much stronger.

3. a, c
A group with lone pairs of electrons (—OH, —NH$_2$) generally activates the benzene ring and causes easier substitution (in this case to the tribromo compound).

4. a, b, e
Primary aromatic and aliphatic amines and amides give nitrogen with 'nitrous acid'.

5. a, d, e
Urea, as the double amide of carbonic acid, behaves as expected of amides with sodium hydroxide solution.

6. b, c, d
Diazotization and subsequent coupling reactions of this type occur only with primary aromatic amines. The products are often useful dyes and indicators (e.g. **c** gives methyl orange).

7. a, d
Benzene, apparently containing three double bonds per molecule, does not show unsaturation with these reagents.

8. d, e
The white precipitate formed on acidification is insoluble benzoic acid.

9. a, e
Acids, amides and acyl chlorides all contain the $>C=O$ group but because of electron rearrangement with the —OH, —NH$_2$ and Cl groups, no condensation reactions occur. These are given only by aldehydes and ketones.

10. a, b, e
These are useful tests for aldehydes, but in some cases one test may not work: e.g. benzaldehyde has no effect on Fehling's solution.

ANSWERS AND COMMENTS

11. a, b, d
—CN attached to a benzene ring behaves as in aliphatic cyanides.

12. a, b, d, e
The lower members of the alcohol and acid homologous series are miscible with water. Esters are usually immiscible, but some lower members have appreciable solubility, e.g. ethyl acetate: 7 g in 100 g of water at 25°C.

13. d
b and d will both give iodoform in test **i**, but the liquid must be a secondary alcohol as shown by test **ii**.

14. c, d
—Cl attached to the benzene ring is not hydrolysed under these conditions.

15. e
1 mole of compound uses 3 moles of sodium hydroxide in hydrolysis. This reaction on stearin **e** is the basis of soap manufacture.

16. b, c, d
The reaction of a diazo group with ethanol is a useful way of removing —NH$_2$ from an aromatic ring.

17. a, b, c, e
Halogen attached directly to the benzene ring is not replaceable by —CN under these conditions.

18. b, c
Because of the carbon-carbon double bond, acrylonitrile can be polymerised into a useful fibre (Acrilan, Courtelle).

19. d
An intramolecular acid anhydride such as [structure: benzene ring with two C=O groups bridged by O, forming phthalic anhydride] will only form when the —COOH groups are on adjacent carbon atoms, for spatial reasons.

20. a, b, c, e
Aluminium triethyl is one of the catalysts used to polymerise ethylene. **b** is one of many very useful Grignard reagents. Lead tetraethyl is the major antiknock additive in petrol. Lithium aluminium hydride is a useful and very powerful reducing agent. As a means of making acetone **d** is of historical importance only.

QUALITATIVE AND QUANTITATIVE ORGANIC ANALYSIS

Answers and comments

1. a, b, c, e
All organic compounds decompose to water and carbon dioxide on heating with copper oxide. Beilstein's copper wire test gives a rapid indication of halogen in a compound.

2. c
Most organic compounds oxidize completely on heating, except salts which leave a deposit of metal carbonate or oxide.

3. a, d
In Middleton's test (or the Lassaigne sodium fusion test) care must be taken if CN^- is present in the filtrate. This forms a white precipitate with $AgNO_3(aq)$ unless removed first by boiling with nitric acid, and thus gives a spurious indication of chloride.

4. e
Chemical purification and drying MUST be done before distillation. In this case $CaCl_2(s)$ is useful as it removes both water and ethanol ($CaCl_2, 6H_2O$ and $CaCl_2, 6C_2H_5OH$ are both formed).
Azeotropes such as the one quoted are quite common.

5. c, d, e
The commonly used inorganic technique of 'leaving copper sulphate solution to crystallize' is unsatisfactory for organic compounds because of the volatility of the solvents. Recrystallizing to constant melting point is the only sure way to purity.

6. b
Most physical properties of straight chain hydrocarbons (and other homologues in other series) show a steady change with increasing molecular weight.

7. a, c, d
In such explosion reactions there is always excess oxygen.

8. a, b, c, d, e
Systematic nomenclature gives one unambiguous name for each compound, and readily affords a way of working out methodically the isomers of hydrocarbons, etc.

9. a, b, c, d, e
Structural formulae often do not describe the compound unambiguously.
In **a** there are two asymmetric carbon atoms, so optical isomers exist. **b** can be arranged with the —COOH groups on opposite sides of the double bond (i.e. *trans*) giving maleic and fumaric acids.
c could be acetamide, so these are functional group isomers.

ANSWERS AND COMMENTS

d could have the chlorine atoms on different carbon atoms.
e could be dimethyl ether, a functional group isomer of ethanol.

10. b
A simple test using sodium is sufficient to distinguish between the two isomers of C_2H_6O. As 1 mole of compound gives $\frac{1}{2}$ mole of hydrogen, there is only 1 replaceable hydrogen atom per molecule.

11. a, b, c, e
Analysis by this (or any other) method does not give the oxygen content of a compound, which must be obtained by difference. The molecular weight could have been measured by a boiling point elevation method, or in other ways.

12. a, b, e
Calculations from results such as these are best done using moles.

13. c, e
Chlorine attached to an alkyl group is not hydrolysed under these conditions.

14. a, b, c, d
This oxygen flask technique is superior to the Carius method of halogen analysis.
The methods described in Questions 11–14 are in wide use, although the quantities needed are usually about 0·1 g of solid to be analysed.

15. b, c, d
All five compounds have the same empirical formula. **b, c, d** and **e** have the same molecular formula but they all have different structural formulae.

16. a, b, c, e
This is a convenient way of finding the number of carboxyl groups in an acid of known molecular weight.

17. a, c, e
The Iodine Number is a useful measure of the degree of unsaturation of oils. These results agree with the oil being the triglyceride of linoleic acid:

$$CH_3-(CH_2)_4-CH=CH-CH_2-CH=CH-(CH_2)_7-COOH$$

Such unsaturated molecules are oxidized fairly rapidly in air and are useful in paint manufacture.

18. d

19. d

20. c
The stages in these sequences can all be performed in the laboratory, although the overall yield of conversion of A to D etc. may be low.

ANSWERS TO REVISION QUESTIONS

Question	1	2	3	4	5	6	7	8	9	10	11	12	13	14	15	16	17	18	19	20
Physical revision 1	a	e	d	e	d	a	a	d	d	e	b	c	c	b	c	c	b	d	d	d
Physical revision 2	c	d	c	d	d	e	c	c	c	a	e	d	d	d	c	a	c	e	e	c
Physical revision 3	b	c	e	d	a	e	c	b	a	d	c	a	d	d	e	d	d	e	c	c
Inorganic revision 1	a	c	d	b	a	e	d	c	e	c	a	e	a	e	a	d	e	c	d	b
Inorganic revision 2	a	e	e	b	e	d	c	d	d	b	d	b	e	b	a	d	a	e	d	c
Inorganic revision 3	b	e	d	c	c	d	b	c	d	d	a	c	d	c	c	e	c	d	e	c
Organic revision 1	e	d	a	b	a	e	e	e	e	e	b	c	e	c	a	d	e	d	a	c
Organic revision 2	e	b	c	a	e	d	c	d	a	e	c	e	e	d	c	a	a	a	b	e
Organic revision 3	e	b	d	c	d	b	d	a	e	d	e	a	d	e	d	d	b	a	b	d

PERIODIC TABLE

(PERIOD)	1M	2M	3T	4T	5T	6T	7T	8T	8T
1	1 **H** (1)								
2	3 **Li** (7)	4 **Be** (9)							
3	11 **Na** (23)	12 **Mg** (24)							
4	19 **K** (39)	20 **Ca** (40)	21 **Sc** (45)	22 **Ti** (48)	23 **V** (51)	24 **Cr** (52)	25 **Mn** (55)	26 **Fe** (56)	27 **Co** (59)
5	37 **Rb** (85)	38 **Sr** (88)	39 **Y** (89)	40 **Zr** (91)	41 **Nb** (93)	42 **Mo** (96)	43 **Tc** (99)	44 **Ru** (101)	45 **Rh** (103)
6	55 **Cs** (133)	56 **Ba** (137)	57 **La** *	72 **Hf** (179)	73 **Ta** (181)	74 **W** (184)	75 **Re** (186)	76 **Os** (190)	77 **Ir** (192)
7	87 **Fr** (223)	88 **Ra** (226)	89 **Ac** **						

TRANSITION METALS (d ENERGY LEVEL FILLING)

LANTHANIDES (4f LEVEL FILLING) *	57 **La** (139)	58 **Ce** (140)	59 **Pr** (141)	60 **Nd** (144)	61 **Pm** (147)	62 **Sm** (150)
ACTINIDES (5f LEVEL FILLING) **	89 **Ac** (227)	90 **Th** (232)	91 **Pa** (231)	92 **U** (238)	93 **Np** (237)	94 **Pu** (242)

RELATIVE ATOMIC MASSES FOR USE IN CALCULATIONS ARE GIVEN IN BRACKETS